MARS,
THE LAST REFUGE
OF HUMANITY

DAVID SANDUA

Mars, the last refuge of humanity.

eBook & Paperback Edition 2023

*"Earth is the cradle of humanity,
but one cannot live in the cradle forever."*

Konstantin Tsiolkovsky

INDEX

I. INTRODUCTION

Overpopulation, depletion of natural resources and global warming are some of the most pressing problems facing the world today. The human race has reached a critical point in its existence, where it not only has to find ways to maintain its resources, but also to ensure its survival. The world's population is growing at an alarming rate and resources are being depleted at an unprecedented rate. Added to this is the threat of global warming, which is rapidly altering fragile ecosystems, causing climate change and natural disasters. All these factors have led humanity to believe that the only way out of this crisis is to explore new frontiers and find a way to colonize other planets. Mars, being the closest planet to Earth and with the potential to harbor life, promises to offer a way out of the looming crisis. Therefore, exploring the possibility of colonizing Mars may be the next step in the evolution of humanity to ensure its survival. This essay aims to explore the possibility of colonizing Mars, its feasibility, and the potential benefits and challenges of making it a reality.

OVERPOPULATION, DEPLETION OF NATURAL RESOURCES AND GLOBAL WARMING

To fully understand the urgency of colonizing Mars, it is important to delve deeper into the problems facing humanity on Earth. The exponential growth of the human population has put enormous pressure on the planet's resources. By 2021, the world's population will exceed 7.9 billion people, more than double that of the 1960s. This population increase has directly contributed to the depletion of natural resources such as water, wood and fossil fuels. The growing demand for energy has resulted in the emission of large quantities of greenhouse gases, leading to global warming, a phenomenon that in itself has a host of deadly consequences. The problem of overpopulation is not new and has been recognized as a global challenge for many decades. The effects of overpopulation can be seen all over the world, from overcrowded cities to increased levels of poverty. Overpopulation puts enormous pressure on the planet's natural resources. It leads to deforestation, which is the large-scale clearing of forests. Deforestation has caused significant damage to the environment, leading to biodiversity depletion, climate change and soil erosion. It also contributes to climate change, as trees function as carbon sinks, absorbing carbon dioxide during photosynthesis and storing it in their tissues. When trees burn or decompose, they release carbon dioxide back into the atmosphere. Another aspect that intensifies the depletion of natural resources is the excessive demand for energy. The use of fossil fuels such as coal, oil and gas as a primary source of energy also

11

has a significant impact on the environment. The burning of these fossil fuels releases large amounts of greenhouse gases into the atmosphere, causing global warming. This is a problem that has been brewing for decades, and its repercussions are being felt in the form of extreme weather conditions, melting of the polar ice caps, rising sea levels and acidification of the oceans. Global warming also has serious consequences for the ecosystem, such as the endangerment of numerous animal and plant species. Global warming is one of the most pressing problems of our time. It is caused by the buildup of greenhouse gases in the atmosphere, primarily carbon dioxide, which traps heat from the sun and drives up global temperatures. The impact of global warming is widespread and dramatic, including more frequent and intense heat waves, droughts, forest fires, storms and floods. Rising sea levels, caused by melting ice caps, are expected to make many coastal areas uninhabitable, and rapidly changing weather patterns will threaten food crops and food security. Rapidly changing weather patterns will threaten food crops and water supplies worldwide. Given the magnitude of these problems, it is clear that humanity needs to find a solution that favors sustainability. The colonization of Mars represents an opportunity to move away from Earth's finite resources and explore alternative and innovative ways to sustain human life. Mars' abundance of resources, such as ice, water and minerals, may offer a solution to Earth's current resource crisis. On Mars, these resources could be used to sustain human life and develop new technologies to aid in the search for alternative energy sources. By providing the opportunity for a fresh start without the confines of Earth's political and economic structures, colonization of Mars offers the possibility of establishing a sustainable society that prioritizes concern for the environment. Overpopulation, depletion of natural resources and global warming are some of the most important problems facing humanity today. With the world's population expected to reach 10 billion by 2050, overpopulation is a problem that will only get worse. Natural resource depletion and global warming are also pressing problems that have detrimental effects on our planet. Although efforts are being made on Earth to achieve sustainability, the colonization of Mars presents an alternative option. Using the resources that Mars offers, we could find solutions to problems that have plagued us for decades. It is time to recognize the urgency of the situation we find ourselves in and take proactive steps to ensure the survival of our species. Colonizing Mars may be one such step.

SOLUTIONS TO ENSURE THE SURVIVAL OF MANKIND

Highlighting the need for solutions to ensure humanity's survival goes beyond the idea of exploring space. It requires understanding the underlying causes of the problems facing humanity and taking deliberate steps to solve them. Overpopulation, natural resource depletion and global warming are the three main challenges that threaten humanity's survival on Earth. The continuous increase in human population has put immense pressure on the planet's resources, leading to their overexploitation and eventual depletion. This has affected the environment, leading to climate change, pollution and loss of biodiversity, all of which have far-reaching consequences for the survival of all living organisms on Earth. The need to find solutions cannot be overemphasized, given the alarming rate at which these problems are worsening. The global community must come together and work to develop sustainable solutions that ensure humanity's survival while preserving the planet's resources for future generations. This means adopting environmentally friendly lifestyles and practices and investing in technologies that reduce our impact on the natural environment. It is also essential to promote education and awareness campaigns that help people understand the importance of preserving the planet's natural resources and reducing their carbon footprint. Only by adopting a holistic approach to solving problems can we offer hope for humanity's survival. Therefore, colonization of Mars could be an additional solution, but it is essential that humanity resolves the factors that threaten Earth's survival before seriously considering colonization.

COLONIZATION OF MARS AS A POSSIBLE SOLUTION

The concept of colonizing Mars as a possible solution to our current problems has been a hotly debated topic in recent times. As mentioned above, humans have been depleting the Earth's resources at an alarming rate, causing an irreversible impact on the planet's ecosystem. The idea of colonizing Mars offers a possible solution to the depletion of natural resources and the problem of population control. Mars, with its abundance of minerals and resources, presents a viable option for a sustainable human colony. The planet also offers much-needed protection from the effects of global warming, such as radiation and extreme weather conditions. The idea of colonizing Mars may seem far-fetched at first, but it holds great promise for ensuring our survival as a species. The possibility of terraforming Mars has been proposed, which involves making the planet habitable by altering its atmosphere and climate. The process would take time and could be complicated, but it presents a long-term solution to the problem of climate change. The concept of colonizing Mars has also sparked renewed interest in space exploration and the development of technology to enable human travel and settlement in space. This could lead to significant advances in a variety of fields, such as medicine, engineering and agriculture, thereby improving the quality of life on Earth. Mars exploration and the concept of colonization offer a promising solution to ensure the survival of our species. The potential benefits could span generations, and it is an idea worth exploring further.

17

Despite the optimism and excitement surrounding the possibility of colonizing Mars, it is important to keep in mind the potential negative consequences and challenges that may arise from such an undertaking. First, the cost of colonizing Mars would be astronomical, requiring immense amounts of resources and manpower. The initial expense of creating habitable space would be enormous, and the maintenance and supply missions would only add to the strain on Earth's already limited resources. Building and operating a Mars colony would require an enormous amount of energy, which could exacerbate global warming and other environmental problems. The colonization of Mars could have significant negative impacts on the planet's fragile ecosystem, as any introduction of terrestrial organisms or materials could have unforeseen consequences for the Martian environment. It is also important to consider the psychological impact of living on a completely new planet, far from the familiarity and comforts of Earth. Humans are social creatures, and the isolation and confinement of a Mars colony could lead to significant mental health problems among the colonists. There is the question of the long-term sustainability and viability of a Martian colony. Despite our efforts to create self-sufficient environments, there is always the risk of technological failure or resource depletion, which could lead to the total collapse of the colony. While the idea of colonizing Mars may seem like a solution to Earth's growing population and resource crisis, it is important to keep in mind the potential negative consequences and challenges that may result from such an undertaking.

II. OVERPOPULATION

Overpopulation is a crucial problem that has intensified over the years. With the human population continuing to grow, it is becoming increasingly difficult to meet people's basic needs. The world's population has grown from just under 2 billion at the beginning of the 20th century to over 7 billion in 2019. This growth has put immense pressure on the planet, especially on its resources, such as water, food and fuel. As a result, environmental degradation, pollution and climate change have become the main effects of overpopulation. The excessive consumption of resources to meet the needs of the growing population has left many people in developing countries in a situation of severe poverty and social inequality. One of the ways in which overpopulation has manifested itself is through the invasion of natural habitats, which has led to the extinction of many plant and animal species. The destruction of forests, urbanization and industrialization are some of the factors contributing to the loss of habitats that support a wide variety of species. This loss of diversity can have important consequences for ecosystems and humans, such as reduced air and water quality, increased rates of disease transmission and ecosystem collapse. The high rate of population growth has led to shortages of land, water and shelter, and these shortages have resulted in competition for resources. The scarcity of resources has led to conflicts and wars between and within nations, further exacerbating the problem. Competition for resources has led to increased immigration from poor or developing countries to developed countries. The depletion of natural resources is another consequence of overpopulation. Fossil fuels, minerals and other resources are finite, and their extraction generates pollution, which has negative impacts on the environment and human health. The growing demand for these resources due to population growth is a major driver of environmental degradation, leading to climate change. The depletion of fossil fuel reserves threatens global economic stability, especially for developing countries. The growth and expansion of developing country economies has increased their dependence on fossil fuels, with the associated challenges of pollution and waste management. Energy is essential for economic growth, and without concerted efforts to transition to sustainable sources such as wind, solar and hydropower, all countries will face imminent depletion of non-renewable energy sources.

The latest impact of overpopulation is global warming. Human activities have been identified as the main cause of global warming. The high concentration of greenhouse gases in the atmosphere is responsible for global warming, and this concentration comes from human activities. Human activities, such as deforestation, industrialization and the burning of fossil fuels, release carbon dioxide and other greenhouse gases into the atmosphere, creating a thick layer that traps the sun's heat and warms the planet. This warming effect is responsible for rising global temperatures, melting glaciers and ice caps, and rising sea levels. Some of the consequences of global warming are the extinction of plant and animal species, the spread of diseases, the destruction of coral reefs, the loss of biodiversity and the increase in natural disasters such as droughts, heat waves, forest fires and floods. Global warming has also disrupted the agricultural sector, causing food shortages and mass migrations.

Given the severity of these challenges, the solution is to colonize

Mars. Although colonization of the red planet poses many challenges, such as radiation exposure, the long-term health effects and huge financial investments are surmountable with the right scientific and technological advances. Colonizing Mars will offer the opportunity to escape the problems arising from terrestrial overpopulation. As space agencies such as NASA and SpaceX accelerate their plans to send humans to Mars, they will also be laying the groundwork for colonization of the red planet. Sustainable settlement on Mars will require the use of innovative technologies and conservation of natural resources. And while it may take decades or even centuries to make Mars a habitable environment, it offers a unique opportunity for humanity to avoid the environmental, social and economic problems caused by overpopulation and depletion of natural resources on Earth. A mission such as the colonization of Mars will inspire the world with the possibilities of science and technology, while contributing to a better understanding of our place in the universe. We must find a solution to overpopulation, depletion of natural resources and global warming to ensure humanity's survival. This mission to Mars may be that solution.

DEFINITION OF OVERPOPULATION

Overpopulation refers to a situation in which the number of people living in a given area exceeds the capacity of the environment to adequately support them. It is a complex problem involving demographic, environmental and socioeconomic factors, among others. Since the beginning of the Industrial Revolution, the human population has grown at an unprecedented rate, resulting in numerous challenges that threaten the sustainability of the environment, humans and other species. Overpopulation interacts with other global problems such as climate change and the depletion of natural resources, forming a vicious circle that aggravates the problems. Overpopulation leads to resource and infrastructure overload, poor living conditions and increased pollution, among other things. It also raises ethical issues regarding the distribution of resources, as some areas may be overpopulated while others may be over-resourced. Various efforts have been made to address overpopulation, such as family planning programs, education and economic development, among others, but the problem persists. Therefore, a comprehensive and integrated approach is needed to ensure that the human population is sustainable and able to achieve a high quality of life without compromising the well-being of other species and the environment. Mars colonization can be seen as a potential solution to overpopulation, but it should not be seen as a substitute for addressing the root causes of overpopulation and associated problems.

CONSEQUENCES OF OVERPOPULATION: LACK OF RESOURCES, POLLUTION AND CLIMATE CHANGE

The consequences of overpopulation are vast and far-reaching, affecting everything from resource availability to climate change. Resource demand is perhaps one of the most immediate and pressing problems posed by population growth. As the number of people in need of food, water and other necessities increases, so does the pressure on the planet's natural resources. This has led to overfishing, deforestation and overuse of freshwater resources. In turn, this has contributed to biodiversity loss, soil degradation and declining air and water quality. Pollution is another major consequence of overpopulation. As more people use fossil fuels and other non-renewable resources, the resulting emissions contribute to air pollution and climate change.

The resulting emissions contribute to air pollution and climate change. Industrialization, urbanization and transportation also emit significant amounts of harmful pollutants, which can have serious consequences for public health. The rise of consumerism and the throwaway culture has also led to an increase in waste production, further aggravating the pollution problem. Overpopulation has played an important role in climate change. As more people consume resources and emit greenhouse gases, the resulting warming trend can cause environmental disasters such as droughts, floods and severe weather events. Climate change also poses a threat to the survival of species, especially those that depend on specific habitats that may be affected by temperature changes. Overpopulation poses a major threat to the ecological health of the planet and the well-being of human society, making the need to explore alternative options, such as the colonization of Mars, ever more pressing.

POSSIBLE SOLUTIONS: SUSTAINABLE LIVING, BIRTH CONTROL AND MIGRATION TO OTHER PLANETS

Sustainable living, birth control and migration to other planets are possible solutions to mitigate the effects of overpopulation. Sustainable living requires less consumption of finite resources such as fossil fuels, water and land by reusing and recycling these resources. It is important to reduce humanity's carbon footprint to limit the effects of global warming in the short and long term. Birth control on the other hand would limit the rate of growth of human populations, slowing the depletion of resources and allowing remaining resources to be sustained over the longer term. It would also promote gender equality and allow women the freedom to choose their reproductive behavior. Migration to other planets, while not a short-term solution considering the challenges of terraforming and the expense of space travel, would serve as a long-term solution for humanity. Exploring other planets and their resources could provide a platform to reduce pressure and competition for resources on Earth and give humanity a chance to survive without the negative effects of global warming, overpopulation and resource depletion. Migration to other planets would not solve the problem of overpopulation on Earth in the foreseeable future. All three solutions suggest the need for a multidisciplinary approach, so that each solution complements the other, thus providing humanity with the opportunity to thrive in a sustainable future. The colonization of

Mars has become an increasingly popular topic in recent years as humanity strives to find solutions to the problems it currently faces. With overpopulation, depletion of natural resources and global warming threatening the survival of our species, the search for options beyond our planet has become not only a possibility, but a necessity. Although some remain skeptical of the idea of life on Mars, recent discoveries suggest that it may be possible to sustain life on the red planet. In fact, colonization of Mars carries with it a number of advantages that could benefit humanity not only in the short term, but for generations to come. One of the biggest advantages of colonizing Mars is the possibility of preserving life beyond Earth. With the current state of the planet, it is increasingly likely that we will run out of resources and face extinction. This is where Mars comes into play as a possible reservoir. Unlike other planets in our solar system, Mars is the most Earth-like in size and composition. It has a day/night cycle similar to ours, a rocky surface, and even once had flowing water. Although the environment is hostile to human life, scientists have already developed a number of technologies that could help us survive there, such as systems that could create breathable air and produce food and water. This means that, should the Earth become uninhabitable at some point, humanity may have the opportunity to start over on Mars. Another advantage of colonizing Mars is the opportunity for scientific exploration and discovery. Mars has long been an object of fascination for astronomers and scientists, and for good reason. Not only does it offer a wealth of information about the history and formation of our solar system, but it also presents a number of unknowns that could be vital to our understanding of the universe and our place in it. By sending more missions, manned or unmanned, to Mars, we could discover more about the planet's geology, its atmosphere and even the possible existence of life. And who knows what other discoveries could be made along the way? Perhaps one of the most compelling reasons to colonize Mars is the impact it could have on our own planet. As we continue to consume resources and produce waste at an alarming rate, the environment is increasingly damaged. Factors such as deforestation, pollution and climate change are already altering our planet irreparably, and we seem to be heading towards a point of no return. Colonizing Mars could relieve the pressure on our planet and give it a chance to recover. We could extract resources and raw materials from the Martian environment, reducing our dependence on Earth's precious resources. And, of course, there is the possibility that we could discover new technologies and life forms that would help us reduce our own impact on the environment. Of course, there are many challenges associated with Mars colonization, many of which are currently unknown. For one, the distance between Mars and Earth poses a huge logistical hurdle. Our current spacecraft are capable of traveling at speeds of about 40,000 kilometers per hour, which means it would take months or even years to reach Mars. And once there, it would be extremely difficult to establish a sustainable environment that could support human life for extended periods of time. The potential health risks associated with life on Mars are currently unknown. This planet has less atmosphere than Earth, which means that radiation levels are much higher. Mars has weaker gravity than Earth, which can lead to loss of muscle and bone mass over time. Despite these challenges, the idea of colonizing Mars remains incredibly attractive. For many, it represents a beacon of hope in an otherwise bleak world. While it may not be the perfect solution to all of our problems, it offers a potential safety net for humanity and the opportunity to explore and discover beyond our own planet. It may also serve as a catalyst for new technologies and ways of thinking that could help us create a more sustainable future. And at a time when the very survival of our species is in question, it may be the only hope we have left.

III. DEPLETION OF NATURAL RESOURCES

The depletion of natural resources is an undeniable threat to the survival of our planet. From forests to oceans, natural resources have been exploited for centuries to sustain civilization.

With the increase in human population and advanced technology, the rate of resource depletion has skyrocketed. One of the most alarming examples is the depletion of fossil fuels such as oil, gas and coal. These non-renewable resources have been extracted without taking into account their finite nature, resulting in their rapid depletion. The production of greenhouse gases emitted by the combustion of these fossil fuels has been one of the main causes of global warming. Another example is the exploitation of forests, which provide essential ecosystem services such as carbon sequestration, water regulation and biodiversity. Deforestation has caused irreversible damage to our environment, causing soil erosion, loss of wildlife habitat and biodiversity, and declining water quality. Exploitation of the oceans for fishing and other resources has resulted in unsustainable practices, causing significant declines in fish stocks and negatively impacting marine ecosystems. The depletion of natural resources not only impacts the environment, but also human lives, affecting both health and the economy. With the depletion of natural resources, access to basic needs such as food, water and shelter is threatened. The depletion of natural resources poses a major and pressing challenge that must be addressed to ensure the survival of our planet and the prosperity of generations to come.

DEFINITION OF NATURAL RESOURCE DEPLETION

Natural resource depletion refers to the process by which the Earth's finite resources are used at a faster rate than they can be replenished. This can occur for a variety of reasons, such as overfishing, deforestation and mining, among others. When natural resources such as timber, oil, water and minerals are extracted from the environment at a rate that exceeds their natural replenishment rate, they become scarce. The depletion of natural resources threatens the survival of several species that depend on these resources for their basic needs, such as food, shelter and drinking water. It is closely related to human population growth and industrialization. As the world's population continues to grow, demand for natural resources increases, placing greater pressure on the planet's resources. Human activities such as the burning of fossil fuels, industrialization and urbanization have led to the depletion of natural resources and environmental degradation. Fossil fuels such as coal, oil and natural gas are finite resources that are depleting much faster than they can be replenished. The depletion of fossil fuels is of great concern, as these resources are a primary source of energy for industrial processes, transportation and electricity generation. In addition to fossil fuels, other natural resources such as fresh water, timber and minerals are also at risk of depletion due to overuse and unsustainable extraction practices. In some regions, overexploitation of resources has led to severe water shortages, soil erosion and deforestation. The depletion of these resources can lead to environmental degradation, loss of biodiversity and ecosystem collapse. The depletion of natural resources has several negative impacts on human society, including the loss of livelihoods for millions of people who depend on natural resources for survival. Fisheries depletion, for example, not only threatens the survival of fish species, but also the livelihoods of fishermen who depend on fishing for their income. Depletion of land resources due to deforestation leads to soil erosion, desertification and loss of agricultural land, which can affect food security and increase the risk of famine. The depletion of natural resources is linked to global warming, which further aggravates environmental degradation. The burning of fossil fuels releases greenhouse gases into the atmosphere, which trap heat and cause the Earth's temperature to rise. Global warming affects weather patterns, sea levels and ecosystems, causing more severe and frequent natural disasters such as droughts, floods, hurricanes and forest fires. The depletion of natural resources is a major concern for the survival of humanity and the planet. Unsustainable use of natural resources is putting immense pressure on the environment and ecosystems, threatening the survival of various species and human society. Human activities such as population growth, urbanization and industrialization are accelerating the depletion of natural resources and urgent action is needed to address this problem. The adoption of sustainable practices such as renewable energy, conservation and efficient use of resources is crucial to avoid the depletion of natural resources and ensure the sustainability of human society.

EFFECTS OF DEPLETION: RESOURCE SCARCITY AND ENVIRONMENTAL DAMAGE

Overpopulation has led to the depletion of the Earth's natural resources, with serious consequences such as resource scarcity and environmental destruction. The world's population has increased exponentially in recent decades, placing an overwhelming demand on the planet's resources. As people consume more food and energy, the resources available to sustain life on Earth have become scarce. For example, the demand for water has increased significantly with population growth, leading to water scarcity and conflicts over water resources. Natural resource depletion has also led to environmental destruction, such as biodiversity loss, deforestation, pollution and climate change. These environmental challenges have caused widespread damage, disproportionately affecting vulnerable communities and exacerbating social inequality. Resource depletion and environmental destruction can have far-reaching consequences for the planet's ecosystems and their ability to sustain life. Biodiversity loss can lead to the loss of important ecosystem services, such as pollination and soil fertility, which are essential for agriculture and food production. Deforestation can lead to soil erosion, causing landslides and reducing the soil's ability to store carbon and regulate the water cycle. Air, water and soil pollution can have serious consequences for human and wildlife health and lead to the loss of ecosystem services. Climate change, caused by the accumulation of greenhouse gases in the atmosphere, can lead to higher temperatures and sea levels, more frequent and severe weather events, and loss of habitats for plant and animal species. These consequences of environmental depletion and destruction are closely interrelated and have complex and long-lasting implications for the future of humanity and the planet.

Resource depletion and environmental destruction also have social, economic and political implications. Conflicts over resources can lead to social unrest, displacement and forced migration. Loss of ecosystem services can reduce agricultural productivity and raise food prices, affecting vulnerable populations dependent on subsistence agriculture. Environmental problems can exacerbate social inequalities and exacerbate poverty, especially in low-income countries, which are most affected by climate change and environmental degradation. The depletion of natural resources and environmental destruction caused by overpopulation pose a major threat to humanity and the planet. The consequences of depletion are complex and far-reaching, affecting not only the environment but also social and economic well-being. The colonization of Mars may offer a solution to some of these challenges; it should not be seen as a substitute for addressing the root causes of environmental depletion and destruction on Earth. Instead, we should focus on promoting sustainable development, reducing greenhouse gas emissions and protecting natural ecosystems to ensure the survival of humanity and the planet.

POSSIBLE SOLUTIONS: CONSERVATION, RENEWABLE ENERGIES AND RESOURCE ALLOCATION

To alleviate the impact of overpopulation and the depletion of the Earth's natural resources, several potential solutions have been proposed, including conservation, renewable energy and resource allocation. Conservation efforts can include reducing waste, promoting sustainable agriculture, and protecting endangered species and habitats. Renewable energy sources, such as wind, solar and hydro, can also help reduce demand for fossil fuels and thus mitigate the effects of climate change. Resource allocation strategies, such as using water more efficiently, implementing recycling programs and regulating population growth, can also be used to address overpopulation and resource depletion. While these solutions may help mitigate the effects of overpopulation and resource depletion on Earth, their effectiveness may be limited, especially in the face of an ever-growing population. Moreover, while hope may rest in the colonization of Mars, it is important to remember that any successful colonization effort will likely require an even greater investment in conservation, renewable energy, and resource allocation in order to ensure the survival and sustainability of the Martian environment. It is clear that addressing overpopulation and resource depletion will require a multi-faceted approach involving individual and collective efforts to reduce waste and increase efficiency, along with more systemic changes in how we manage and allocate our natural resources. The colonization of Mars has attracted attention as a possible solution to the growing problems of overpopulation, natural resource depletion and global warming. It offers the opportunity to create a new habitable environment in which humans could thrive while preserving planet Earth for future generations. The idea of colonizing Mars has been around for several decades but has gained momentum in recent years due to technological advances and growing concern about the sustainability of life on Earth. In the race to colonize Mars, there are several key challenges that must be addressed. One of the most important is the limited availability of resources needed to sustain human life, such as food, water and energy. Mars has no breathable atmosphere and its soil is toxic to plants, making it difficult to grow crops. Therefore, it will be necessary to develop methods to produce food and obtain water, possibly through the use of hydroponics, water reclamation and terraforming technologies. Energy is also a critical resource, and sustainable methods will need to be developed to produce it, for example, through solar energy. Another major challenge is the harsh environmental conditions on Mars, such as extreme temperatures, radiation exposure and dust storms. Therefore, new technologies will need to be developed to protect humans from these conditions, such as creating radiation-proof structures that can withstand the harsh Martian environment. Any human colonization of Mars must be carried out in a sustainable and ethical manner, ensuring that it does not harm the Martian environment or its potential inhabitants. The colonization of Mars will require careful planning and coordination by international organizations, governments and private companies to ensure its success and the well-being of its inhabitants. Despite these challenges, the potential benefits of Mars colonization are immense, including the expansion of human knowledge, ensuring the survival of our species, and leading to technological advances that can benefit us on Earth. It is imperative that we pursue this endeavor with the utmost care and responsibility. The colonization of Mars represents a unique opportunity to address some of humanity's most pressing concerns as we embark on a new frontier of exploration and discovery.

IV. GLOBAL WARMING

Global warming, also known as climate change, is one of humanity's most pressing concerns. It is the gradual increase in the average temperature of the Earth's surface caused by greenhouse gases. These gases, such as carbon dioxide, methane and nitrous oxide, trap heat in the atmosphere, causing an overall warming effect. Although some may argue that global warming is a natural phenomenon, it is difficult to ignore the scientific evidence that points to human activities as the main contributor to the increase of these greenhouse gases in the atmosphere. Burning fossil fuels for energy production, transportation, deforestation and intensification of agriculture are some of the human activities that contribute to the increase of greenhouse gases in the atmosphere. If left unchecked, global warming will have adverse environmental effects, such as sea level rise, ocean acidification, more frequent and severe weather conditions, and the extinction of several species. The consequences of these impacts will be far-reaching and devastating. However, there is still hope that humanity can mitigate the effects of global warming through a combination of initiatives, such as the transition to cleaner and renewable energy sources, the reduction of greenhouse gas emissions, reforestation and the adoption of sustainable lifestyles. These efforts must be made on a global scale to achieve significant progress towards a more sustainable future.

DEFINITION OF GLOBAL WARMING

Global warming refers to the long-term increase in the average temperature of the Earth's surface, mainly due to the accumulation of greenhouse gases in the atmosphere. These gases, such as carbon dioxide, methane and nitrous oxide, trap the sun's heat and prevent it from returning to space. Human activities, such as the burning of fossil fuels, deforestation and industrial processes, have dramatically increased the concentration of these greenhouse gases in the atmosphere, causing global temperatures to rise. This increase in temperature can have far-reaching impacts on the Earth's climate, such as melting of the polar ice caps, rising sea levels, more frequent and severe weather events, and changes in precipitation patterns. Although there is still some debate about the exact magnitude of the effect humans are having on global warming, the overwhelming scientific consensus is that it is occurring and that it poses a significant threat to the planet. Many solutions have been proposed to help mitigate the effects of global warming, such as reducing our dependence on fossil fuels, reducing emissions from agriculture, and using carbon capture and storage technologies. As temperatures continue to rise and the effects become more severe, it is becoming increasingly urgent to take more aggressive action to address this problem.

GLOBAL WARMING: RISING SEA LEVELS, EXTREME WEATHER EVENTS AND DISRUPTION OF ECOSYSTEMS

Global warming continues to cause a variety of consequences, from sea level rise to extreme weather events and ecosystem disruption. Sea level rise due to melting glaciers and ice sheets is one of the most important effects of global warming on the world's coastal regions. Scientists predict that by the end of the century, seas could rise by up to one meter, which would have a significant impact on coastal communities. Coastal erosion, flooding and saltwater intrusion into freshwater sources are just some of the problems expected to be exacerbated by rising sea levels. Global warming is intensifying extreme weather events such as hurricanes, typhoons, droughts, floods and heat waves, resulting in devastating social, economic and environmental impacts. In 2018, natural disasters caused by climate change triggered $155 billion in losses worldwide, up from $143 billion in 2017. Also associated with disruptions to ecosystems, such as deforestation, coral reef bleaching, biodiversity loss and algal blooms. Climate change is causing species migration, which is reorganizing ecosystems around the world and threatening the survival of many species. These effects, impact food sources, water and human security. The consequences of global warming are wide-ranging and affect the social, economic and environmental spheres. Communities, governments and individuals must take immediate action to mitigate these impacts.

POSSIBLE SOLUTIONS: REDUCTION OF CARBON EMISSIONS, USE OF CLEAN ENERGY AND CARBON SEQUESTRATION TECHNOLOGY

The current climate crisis is undoubtedly one of the greatest threats facing humanity today. The Earth's temperature has already risen by nearly 1 °C and, if the current rate of carbon emissions is maintained, it is projected to increase by up to 3.2 °C by the end of this century. To avoid catastrophic consequences, it is essential to reduce greenhouse gas emissions. Carbon capture technology is one of the many tools available to help us achieve this goal. It involves capturing carbon dioxide from various industrial processes, such as power plants, and storing it in subway reservoirs. Although this technology is still in its early stages and faces technical and economic challenges, it has the potential to play an important role in reducing carbon emissions. Another promising approach is to turn to clean energy sources, such as solar, wind and geothermal, to replace the fossil fuels that currently dominate our energy mix. To achieve this transition to clean energy, governments around the world must undertake bold policy initiatives, such as introducing carbon taxes and investing in research and development of new clean energy technologies. There must be a significant reduction in global carbon emissions. This can be achieved by drastically reducing the frequency with which individuals use personal automobiles, encouraging the use of environmentally friendly alternative modes of transport and the adoption of greener lifestyles.

The fight against climate change requires a collective effort from all of us. By adopting these solutions, we can make planet Earth a more sustainable and livable place for ourselves and future generations. As humanity's impact on Earth becomes increasingly evident, many are turning to the idea of colonizing Mars as a solution to the problems of overpopulation, natural resource depletion and global warming. While it may sound like science fiction, there are many reasons why Mars could be a viable option for future human settlement. For starters, Mars has a number of similarities to Earth that make it a more hospitable environment than other potential locations, such as the Moon. For example, Mars has a slightly denser atmosphere, which could provide some protection from harmful radiation bombarding the planet's surface. Mars also has a day/night cycle similar to Earth's, which could help regulate the sleep cycles of colonists. Mars has abundant water in the form of ice, which could be used for drinking, growing crops, and creating rocket fuel. Another reason Mars is an attractive option for colonization is that it has a number of resources that could be mined and used to support a human settlement. Mars is rich in iron, which could be used to construct buildings and infrastructure, as well as precious metals such as gold and platinum, which could be mined and traded. Mars also has a significant amount of carbon dioxide in its atmosphere, which could be converted into breathable oxygen and used to grow plants. Of course, the challenge of colonizing Mars is not insignificant. The journey to Mars itself would be a major hurdle, as it takes about six months to travel from Earth to Mars with today's technology. Once there, the colonists would have to build a self-sufficient community that could survive in the harsh Martian environment. This would require the development of advanced life support systems, such as food production, waste management and air recycling. Colonists would also have to prepare for Mars' extreme weather conditions, such as dust storms, extreme temperatures and low atmospheric pressure. Despite these difficulties, the idea of colonizing Mars has gained momentum in recent years, and both private companies and national governments have announced plans for missions to the Red Planet. In 2020, NASA launched the Mars Perseverance Rover, which will search for signs of ancient life and collect samples that could be returned to Earth for further study. The Mars 2020 mission also includes a small, unmanned helicopter that will attempt to fly in the thin Martian atmosphere, demonstrating the feasibility of powered flight on another planet. Private companies such as SpaceX, founded by Elon Musk, are also taking important steps in the development of spacecraft and technologies that could enable human colonization of Mars. SpaceX has announced plans to send a manned mission to Mars as early as 2024, using its Starship spacecraft and Super Heavy rocket. The company's ultimate goal is to establish a self-sustaining city on Mars, with thousands of inhabitants living permanently on the planet. While the idea of colonizing Mars is undeniably exciting, there are also valid questions about the ethics and feasibility of such an undertaking. Some have argued that focusing on Mars diverts attention from efforts to solve Earth's problems, such as poverty, inequality and environmental degradation. Others have expressed concern about the potential ecological impact of introducing human life to a pristine environment such as Mars, and the possibility of contaminating the planet with terrestrial microbes that could complicate future scientific exploration. Despite these concerns, the prospect of colonizing Mars remains a potent symbol of human ambition and ingenuity. It is a vision of a future in which humans not only survive but thrive in a new world beyond our own. Whether this vision becomes a reality remains to be seen, but there is no doubt that the exploration of space and the search for new frontiers will continue to captivate and inspire us for generations to come.

V. COLONIZING MARS AS A POSSIBLE SOLUTION

One possible solution to the dire global environmental situation is the colonization of Mars. Although it may seem a far-fetched and unrealistic idea, the concept of Mars colonization has been studied and researched extensively by scientists and engineers for many years. The idea of starting a new civilization on Mars offers numerous advantages, such as the creation of a self-sufficient and independent society, the discovery of new scientific knowledge, and the possibility of expanding human knowledge of the universe. The colonization of Mars could serve as a safeguard against possible terrestrial extinction events, such as global pandemics or meteorite impacts. As a possible solution to the challenges of overpopulation, natural resource depletion and global warming, Mars colonization could offer new hope for the future of humanity. At first glance, the idea of colonizing Mars may seem like a far-fetched dream, but the concept has been around for many years. NASA's Mars Exploration Program has been researching and exploring Mars for decades, with the ultimate goal of eventually sending humans to the planet. Many other private companies, such as SpaceX and Blue Origin, have also expressed interest in colonizing Mars in the future. Thanks to extensive research and technological advances, we have come to better understand the planet's environment and how it could potentially serve as a new home for humanity.

One of the most significant advantages of colonizing Mars is the possibility of creating a self-sufficient and independent society.

By developing the necessary infrastructure and systems, such as agriculture, water and air production, and power generation, a Mars colony could become fully self-sufficient. This would allow the establishment of a stable society that would not depend on Earth's limited resources and could function independently of any possible catastrophes on Earth. The creation of a self-sufficient society on Mars could serve as a model of how humanity can live sustainably and independently, potentially inspiring similar developments on Earth. In addition to serving as a potential model for sustainable life, colonization of Mars could offer great opportunities for scientific discovery and exploration. The planet's unique environment offers many research opportunities in fields such as geology, meteorology, and astrobiology. The study of Mars could also provide valuable insights into the formation and evolution of the solar system, expanding our understanding of the mysteries of the universe. The colonization of Mars could ultimately lead to the creation of new technologies and innovations that could have potential applications on Earth. Perhaps the most significant advantage of Mars colonization is the role it could play in protecting humanity from potential extinction episodes. Earth has suffered numerous catastrophes throughout its history, from pandemics and natural disasters to asteroid impacts. These events could pose a significant risk to humanity's survival, making it essential to establish an independent and self-sufficient society off-planet. Mars offers a viable option for establishing a new civilization in case something catastrophic happens to Earth. The establishment of a colony on Mars could allow the creation of a worldwide network of colonies that would ensure the survival of humanity in the event of a global catastrophe. Although Mars colonization offers many opportunities, it is not without its challenges. The planet has a harsh environment, with dangerous levels of radiation, intense dust storms and extreme temperature fluctuations. Establishing a colony on Mars would require overcoming significant technological and engineering challenges, from developing a suitable spacecraft to building a sustainable, habitable colony on the planet's surface. The cost of such an undertaking would be immense, with estimates ranging from tens of billions to trillions of dollars. Despite these challenges, the potential benefits of Mars colonization make it a crucial consideration for the future of humanity. The exponential growth of the world's population, coupled with the increasing depletion of the Earth's resources and the catastrophic effects of global warming, have made it essential to consider alternative solutions. While there are many potential solutions to these challenges, the colonization of Mars offers a unique opportunity to create a new, fully self-sufficient and independent civilization. The scientific knowledge and technological innovations that could result from such an endeavor would have important implications for the future of humanity. The idea of colonizing Mars may sound like a fantasy, but it is an idea that has been researched and considered by scientists and engineers for many years. The establishment of a self-sufficient, independent society on Mars could serve as a model for sustainable living and have important implications for advancing our knowledge of the universe. The role that Mars colonization could play in protecting humanity from catastrophes should not be underestimated. Although Mars colonization poses significant challenges, given the current environmental challenges and resource depletion we face, it remains a crucial option for ensuring humanity's survival.

STRATEGIES AGAINST OVERPOPULATION, RESOURCE DEPLETION AND GLOBAL WARMING

The potential for colonization of Mars could address the consequences of overpopulation, natural resource depletion and global warming. Overpopulation is a critical global problem that is increasing at an alarming rate. According to the United Nations, the world's population is expected to reach 9.7 billion by 2050, putting enormous pressure on the planet's resources. The colonization of Mars could be a solution to this problem, as it offers a new habitat for humans to live and work. The vast expanse of the planet will provide enough space to accommodate Earth's growing population. Colonizing Mars could also offer new territories and resources needed to sustain human life in the long term. Human colonies would be able to harness and sustainably utilize Martian resources such as water, metals and minerals. This, in turn, would decrease Earth's dependence on scarce and finite resources, which are becoming increasingly depleted. Natural resource depletion is another critical issue that is affecting the sustainability of the planet. Overpopulation, urbanization and industrialization are causing immense damage to the Earth's ecosystems. Climate change experts argue that urgent action is imperative to ensure the sustainability of the planet. Mars colonization is one way to address this problem. The use of Martian resources would allow humans to decrease their dependence on Earth's natural resources, reducing pressure on the planet's ecosystems. The planet's resources could be used sustainably to minimize environmental degradation, ultimately strengthening long-term ecological functioning. The colonization of Mars may also help address global warming, which has become a major concern. The burning of fossil fuels is believed to be the main cause of climate change. As humans attempt to reduce their dependence on these fuels, transitioning to renewable and clean energy sources is imperative. The colonization of Mars could contribute to this process by providing the opportunity to explore new energy sources, such as solar, nuclear and geothermal. The planet's atmosphere and climate present new challenges and unique opportunities for energy harnessing. The development and application of this technology, which would be necessary to colonize Mars, could also lead to important technological advances that could benefit efforts to address climate change on Earth. Mars colonization could support scientific research, technological advances and space exploration. Mars colonization could support scientific investigation of the universe. Scientists could investigate Mars scientifically and understand the complex chemical processes and reactions that are harmful to humans. By understanding these reactions, scientists can develop new techniques to counteract or reduce their impact on human health. In the long term, this could lead to the development of new tools and technological innovations that could be used to benefit human life. It is hoped that the colonization of Mars could trigger technological advances that teach scientists about space and the universe, providing valuable information for scientists studying the Earth and the environment. The development of the technology needed for Mars colonization may also lead to the innovation of new products and services, which can benefit society in a variety of ways. Colonizing Mars could provide an opportunity to unite humanity in an extraordinary way. The colonization of Mars casts a futuristic light and proposes the dream of humans living on another planet. Colonization is a long and extensive process that involves people from different backgrounds and disciplines working together. This would require international collaborative efforts that could unite humanity as never seen before. Humans have long been fascinated by the idea of space and the universe, and Mars would offer a new opportunity for people to work together, learn from each other, and explore the unknown. By coming together to undertake this project, it is possible to bring together and share ideas from diverse fields and communities, fostering creative problem solving and benefiting the human race for generations to come.

The colonization of Mars holds immense potential, such as offsetting the consequences of overpopulation, depletion of natural resources and global warming. The colonization of Mars would provide humanity with a new habitat, essential for the long-term survival of our species. It presents a comprehensive means of learning about the universe, science, technology and advanced resources that could lead to further advances in innovation, tools and services. Although the colonization of Mars is still in its early stages, it provides hope and a vision of a prosperous future for humanity.

HISTORY OF MARS EXPLORATION AND COLONIZATION POTENTIAL

The history of Mars exploration dates back to the 17th century, when a Dutch astronomer, Christiaan Huygens, discovered the existence of the planet through his telescope. It was not until the 1960s that mankind made significant advances in the exploration of the planet. The Soviet Union launched multiple probes to Mars, and its Mariner 4 became the first spacecraft to capture close-up images of the planet's surface in 1964. The United States also launched several probes, with Viking 1 being the first to successfully land on the surface of Mars in 1976. Over the years, other missions have been conducted with the goal of obtaining more information about the geology, atmosphere and possibility of life on Mars. The most recent and most prominent is Perseverance, NASA's Mars rover, which landed on the planet's surface in February 2021 and is currently exploring Jezero Crater for signs of past microbial life. The possibility of colonizing Mars has been a topic of interest among researchers and science enthusiasts for decades. Although the Red Planet presents numerous challenges that must be overcome before colonization is possible, such as its harsh environment, thin atmosphere, and lack of oxygen and water, the potential benefits are enormous. For one thing, Mars has a 24.6-hour day and is slightly farther from the Sun than Earth, giving it an Earth-like climate. Mars has vast natural resources, such as iron, aluminum and titanium, which can be used to build infrastructure. Moreover, the colonization of Mars would also serve as a backup plan for Earth in the event of a catastrophe, such as a massive asteroid impact or the eruption of a supervolcano, which could make Earth uninhabitable. NASA's Mars exploration program, along with those of other nations, is not only focused on finding ways to sustain human life on the planet, but also on understanding Mars as a planet and as a potential source of useful resources for Earth. According to NASA, technology developed to explore and live on Mars could also be used to address challenges on Earth, from developing new and efficient energy sources to advancing sustainable agricultural development in hostile environments. The benefits of space exploration do not stop with the discovery and colonization of another planet, but also with the technologies and innovations that are developed in the process. Despite the potential difficulties of colonizing Mars, such as its thin atmosphere and lack of oxygen and water, scientists and researchers are working on ways to overcome these challenges. NASA's Mars 2020 mission, which launched Perseverance, also included a demonstration of a device called Moxie, which converts carbon dioxide from Mars' atmosphere into oxygen. This technology could be used in the future to terraform the planet, creating an environment more Earth-like and more suitable for human life. Much research has been done on creating sustainable habitats on Mars, potentially using 3D printing technology to build structures and using local resources, such as Martian soil, for building materials. The idea of creating a self-sustaining colony on Mars is not new, and numerous organizations have developed proposals on how to achieve it. SpaceX, founded by entrepreneur Elon Musk, has been one of the biggest proponents of Mars colonization. Musk claims that humanity must become a "multiplanetary species" if we are to ensure our survival. SpaceX's current plan is to use its Starship spacecraft, which has yet to be fully developed and tested, to transport people and resources to and from Mars. The company aims to have a sustainable colony of up to one million people on the planet by 2050. This plan is not without controversy, with critics pointing to the significant costs and risks involved, as well as concerns about the potential environmental impact on the planet. Overall, the history of Mars exploration has demonstrated humanity's fascination with the Red Planet and its potential for future colonization. Although there are numerous challenges to overcome before establishing a sustainable colony on Mars, the potential benefits are enormous, both in terms of scientific research and the possibility of ensuring humanity's survival in the event of catastrophes on Earth. The history of space exploration has shown that the possibilities for new discoveries and innovations are endless, and it is therefore important that we continue to push the boundaries of our understanding of the universe.

EARTH vs. MARS: CHARACTERISTICS AND CHALLENGES OF EXPLORATION

When it comes to exploring and colonizing Mars, one of the most critical factors to consider is the differences between Earth and Mars. In many ways, Mars is similar to Earth, but it also presents some unique challenges that must be addressed. One of the main similarities is that both planets have days and nights, and their rotation around the sun lasts about the same length. Mars is much smaller than Earth, with a diameter only half that of our planet. This smaller size means that it has a weaker gravitational pull, which in turn affects its atmosphere and its ability to retain heat and water. Another important difference is that Mars has a much thinner atmosphere than Earth, with less than 1% of the atmospheric pressure at sea level on our planet. This thin atmosphere means that Mars lacks the protective ozone layer that absorbs the sun's harmful ultraviolet rays. Without this protection, astronauts on Mars would be much more vulnerable to radiation exposure, which could have serious health consequences. The thin atmosphere makes breathing difficult, as it lacks the oxygen that is abundant on Earth. Mars also has a different climate than Earth, with much colder temperatures and more extreme weather patterns. The average temperature on Mars hovers around -80 degrees Fahrenheit, much colder than anywhere on Earth. This extreme cold is partly due to Mars' distance from the Sun, as it is much farther away than Earth, but it is also affected by the thin atmosphere and lack of a magnetic field to protect it from the solar wind. This harsh climate presents a unique set of challenges for any mission that intends to explore or colonize Mars. Despite these challenges, there are several similarities between Earth and Mars that could make it an attractive destination for human exploration. One of the most intriguing is the presence of water on Mars, confirmed by several missions. Although the water on Mars is not in the form of liquid oceans as on Earth, there are indications of liquid water beneath the surface, as well as ice sheets at the poles. This water is essential for life and could be used to sustain human habitation on the planet. Another similarity is the presence of minerals and other resources on Mars, such as iron, magnesium, and silicon. These resources could be used to build and maintain human settlements on the planet, as well as to fuel future missions to Mars and beyond. The presence of these resources also raises the possibility of establishing a self-sustaining colony on Mars, which could help ensure the survival of the human species in the event of a catastrophe on Earth. Despite these similarities, there are several challenges that must be addressed before humans can successfully explore and colonize Mars. One of the main challenges is the distance between the two planets, which means that any mission to Mars would require a long journey through space. This distance would also make communication with Earth difficult, as messages sent to and from the planet would be significantly delayed. Another challenge is the need to develop the technology and infrastructure necessary to sustain human life on the planet. This includes developing systems to produce oxygen, water and food, as well as to generate electricity and maintain a comfortable temperature within habitats. Systems would be needed to manage waste and limited resources such as water and air. There are the health risks associated with long-duration spaceflight and living on a planet like Mars. Exposure to radiation and other risks associated with space travel could have long-term effects on astronaut health, while the harsh climate and limited resources on Mars could pose challenges to maintaining human health and well-being. Addressing these challenges will require significant resources, expertise and collaboration from the global community, as well as a sustained commitment to exploration and innovation. Although Mars presents some unique challenges for human exploration and colonization, it also offers a number of opportunities for scientific discovery, resource utilization, and long-term survival of the human species. Understanding the similarities and differences between Earth and Mars is essential to developing effective strategies for exploring the planet and establishing a sustainable human presence there. Overcoming the challenges associated with Mars exploration will require collaboration, innovation, and sustained commitment from individuals, governments, and organizations around the world. The idea of colonizing Mars has been gaining attention as a solution to the challenges facing humanity on Earth, such as overpopulation, depletion of natural resources and global warming. As Earth's resources are finite and the population continues to grow, it is becoming increasingly urgent to search for alternative habitats. Mars, which is about the same size as Earth and has some similarities in terms of resources and atmosphere, has been identified as a possible destination for human colonization. Several private companies such as SpaceX and government agencies such as NASA have been exploring this possibility. The Red Planet is no cakewalk. It is a harsh and unforgiving environment, presenting numerous challenges that must be addressed before we can begin colonizing it. These challenges range from the psychological aspects of living in a strange land to the technical difficulties of designing space habitats to the development of sustainable agriculture and resource management systems. Perhaps one of the most significant challenges of Mars colonization is the psychological impact of living in an isolated and harsh environment for long periods of time. Humans have evolved to thrive in the terrestrial environment, with pleasant temperatures, diverse ecosystems, and an atmosphere that provides us with the right mix of gases to breathe. On Mars, conditions are very different, with a thin atmosphere and radiation that could pose a danger to human health. The psychological impact of isolation in this environment could lead to depression, anxiety and other mental health problems. As a result, astronauts on Mars missions would have to undergo rigorous psychological testing and training to develop coping mechanisms for life in a strange and isolated environment. Social and recreational activities would also have to be designed to help astronauts maintain morale and a sense of connection to their home planet. The design of space habitats is another major challenge that must be addressed in the colonization of Mars. These habitats must be designed to provide a protective shield against radiation, temperature fluctuations, and atmospheric conditions on the planet. Materials that have proven effective against radiation on Earth may not be as effective on Mars, so new materials and designs will have to be developed. Habitat power and energy systems must be designed to be self-sufficient and sustainable. The existence of solar power on Mars makes it an ideal candidate for harnessing the sun's energy, but this will require the development of new storage and distribution systems to make use of the energy when sunlight is not available.

The development of sustainable agriculture and resource management systems is also a key challenge in the colonization of Mars. Food and water will be scarce and must be conserved. The creation of a closed-loop nutrient and water recycling system would be critical for sustainable agriculture on Mars. This would require processing and reuse of astronaut waste. Agro-ecosystems could also be developed using technologies such as hydroponics and aeroponics, which allow cultivation with minimal inputs. This approach would make it possible to grow crops in Mars soil, which is toxic to humans. Manipulating the Martian environment to facilitate agriculture could have significant unintended ecological consequences, and great care must be taken to ensure that the Martian environment is not irrevocably damaged. Overpopulation, depletion of natural resources and global warming have led humanity to desperately seek solutions to ensure its survival. Hope rests in the colonization of Mars.

Mars colonization is seen as one way to address these pressing challenges. The challenges facing Mars colonization are immense. Astronauts will have to cope with the psychological impact of living in an isolated and harsh environment for extended periods. They will also have to deal with the technical difficulties of designing space habitats and providing a self-sufficient and sustainable energy system. Sustainable agriculture will have to be developed to ensure food security on the planet. Despite these challenges, the prospect of colonizing Mars is attractive and we must press ahead with the scientific and technical innovations that will make this dream a reality. Colonizing Mars could help us better understand our place in the universe and help us forge a sustainable future here on Earth.

VI. BENEFITS OF COLONIZING MARS

Mars colonization provides numerous benefits, both practical and aspirational. From our practical point of view, Mars serves as an alternative survival option in the event of global catastrophe; it also allows us to further explore the cosmos and expand our interstellar presence, learning more about our own planet and the universe as a whole. Scientists and researchers studying Mars could gain vital insights into the processes and history of our solar system. A recent study in the International Journal of Astrobiology analyzed the geological history of Mars and concluded that the planet had the potential to harbor microbial life. This would open an avenue for future research into extraterrestrial life. The colonization of Mars also has practical advantages beyond the purely scientific aspect. The technological advances needed for such a mission could be harnessed for use on Earth, contributing to sustainability efforts and reducing dependence on non-renewable resources. This includes improvements in energy, transportation, waste management and other utilities. The mission would require the development of new materials, manufacturing processes and medical technologies that could be used to improve life on Earth. The Martian environment offers the opportunity to develop unique and autonomous economies and cultures, with the emergence of new markets, social structures and scientific communities. Mars' lower gravity and different atmospheric composition pose unique challenges for the agricultural, mining, and construction industries, encouraging the development of new technologies and techniques. With the potential for terraforming, Mars could become a self-sufficient home for humanity, free from the resource constraints and geo-political tensions that plague Earth. The colonization of Mars offers a unique opportunity to unite humanity under a common flag. Instead of competing for our planet's resources and territory, we could cooperate to build a new future for ourselves in space. This would require international collaboration on an unprecedented scale, opening up the possibility of cooperation and peace in a world where conflict and division have been rife. Working together toward a common goal, humanity could transcend the limitations that have prevented us from realizing our full potential. All of these benefits come at a considerable cost. The current state of technology offers limited capability to reach Mars: the trip would take approximately eight months, even with the fastest rocket engines, and the landing process is incredibly complex and requires sophisticated equipment. Such an undertaking would require a massive mobilization of resources, as well as investment from multiple governments, private companies and philanthropic organizations. The environmental and social impacts of such a mission would also need to be taken into account, ensuring that the risks involved are not outweighed by the rewards involved. However, the benefits of colonizing Mars are too valuable to ignore. The future of humanity depends on our ability to expand our horizons and seek new frontiers. This is not simply science fiction or idle curiosity. Humanity faces a very real and pressing threat: overpopulation, resource depletion and climate change. We need to address these problems urgently and the time to start is now. Colonizing Mars may seem like a distant dream, but as we have seen throughout history, the impossible becomes possible when we set our minds to it.

Colonizing Mars is not just a fantasy; it is a necessary

undertaking if we are to ensure the survival of our species and expand our understanding of the universe. The benefits are wide and varied, from scientific exploration to economic and social development to international cooperation. Although the challenges are daunting, we have the tools and the ingenuity to overcome them. As we face the formidable challenges ahead, let us take comfort in knowing that we are capable of great things and that our destiny lies beyond the horizon.

BENEFITS, ABUNDANCE OF RESOURCES, SUSTAINABILITY AND SCIENTIFIC ADVANCES

One of the potential benefits of colonizing Mars is that it will allow the exploitation of abundant resources that are not readily available on Earth. Mars is rich in minerals such as iron, titanium and silicon, which are essential components of the technology and building materials that humans depend on. Mars has large amounts of water in the form of ice, which can be used for drinking, farming and industrial processes. Colonization of Mars could ensure a stable source of materials essential for human life.

The colonization of Mars could promote sustainability on Earth by reducing pressure on our planet's resources by reducing the exploitation of non-renewable resources on Earth. Another potential benefit of Mars colonization is that it provides a new frontier for scientific discovery and technological advancement. By exploring Mars and studying its geology, atmosphere, and history, scientists can better understand the origins of the universe and lay the groundwork for deep space exploration. Colonization of Mars would require technologically advanced systems to support human life, driving technological advances that could have a significant impact on Earth and help solve major challenges such as climate change and energy sustainability. For example, advances in sustainable energy technology, such as solar power and nuclear fusion, could be derived from sustaining life on Mars. Overall, the colonization of Mars may bring about a significant change in the way humans approach resource management and technological advancement, with the potential to reduce some of the most pressing challenges facing humanity today. To effectively harness the benefits of Mars colonization, colonization plans must focus on sustainability, environmental preservation, and responsible use of resources to ensure the health of both Mars and Earth. Sustainability would involve the use of renewable energy sources, such as solar, wind and hydro-electric, to power life on Mars. Such practices would reduce the planet's carbon footprint and provide an opportunity for innovation through the creation of new technologies needed to store, manage and transport excess energy. Preserving Mars as a pristine environment would be essential to ensure the continued efficiency of the planet's resources. The first human inhabitants will need to establish strict environmental guidelines and avoid unnecessary disturbances that could damage Mars' natural ecosystems. This involves designing habitats and infrastructure that minimize negative human impact. Robotic missions could be used to deploy critical infrastructure to ensure the preservation of the planet's natural environment while allowing humans to explore and conduct scientific research on Mars. Prioritizing responsible resource consumption would ensure that the planet's riches are used efficiently to meet the colony's needs while avoiding waste and environmental damage. With Mars' limited resources, resource exploitation must be optimized and managed in a sustainable manner that minimizes depletion of non-renewable resources or environmental damage. Mars colonization holds great promise in terms of resource abundance, sustainability and scientific advances. To realize such benefits, Mars colonization must focus on environmental preservation and responsible resource management. By prioritizing these principles, colonization could reduce the footprints of overpopulation,

natural resource depletion and global warming, and help humanity find effective solutions to ensure survival and sustainability in the years to come.

EVALUATION OF ECONOMIC, POLITICAL AND SOCIAL BENEFITS

The colonization of Mars can bring numerous economic, political and social benefits. Economically, Mars offers great opportunities in resource extraction and commercial activities that can generate substantial revenues for the colonies and the home planet. Mars possesses abundant reserves of minerals such as iron, nickel, cobalt, and magnesium, which can be used to build a self-sufficient economy on the planet. Colonization efforts require the development and deployment of advanced technologies, which can stimulate technological innovation, research and development. As a result, Mars colonization can create a new space economy that unlocks humanity's growth potential, allowing us to reach greater heights. On a political level, the colonization of Mars has the potential to unite humanity around a common goal, the dream of colonizing a new world. The establishment of colonies will require the collaboration of multiple nations, institutions and private entities, leading to the formation of a global coalition. The international cooperation necessary for Mars colonization can boost diplomacy and collaboration between nations and inspire future collaborations in science and innovation that can benefit humanity. Mars colonization may encourage the development of new political systems and governance structures that could solve some of the problems of our current political systems. Mars colonies can serve as an experiment to test different political paradigms, which could lead to the creation of novel political systems that are more responsive to people's needs. This would serve as a valuable lesson for future political experimentation on Earth. From a societal point of view, Mars colonization can bring a number of long-term benefits that enhance human security through improved knowledge, technology and innovation. Colonization of a new world requires the development and use of advanced technologies, which can have far-reaching benefits in a variety of fields, such as aerospace, medicine, engineering, and materials science. These innovations and technological advances can accelerate humanity's progress and solve the problems of overpopulation, natural resource depletion and global warming that threaten its survival. The establishment of a human colony on Mars may spur the formation of a new culture distinct from existing ones. The new environment of Mars may lead to the formation of unique traditions, customs, arts and literature, which could eventually be adopted by the home planet and become part of our world heritage. More importantly, the colonization of Mars can act as a beacon of hope for humanity, inspiring people to work toward a common goal and providing a sense of purpose and direction. After all, this uplifting ideal of colonizing a new planet, where humanity can start anew, presents a vision of a brighter and more optimistic future, a future worth fighting for. The colonization of Mars carries significant economic, political and social benefits that could transform the destiny of humanity. The establishment of a human colony on Mars is not just a technological feat, but a visionary endeavor that embodies our aspiration to rise above our earthly limitations and create a better future for ourselves and generations to come. We must continue to invest in the research and development needed to make Mars colonization possible and turn our dreams into eventual reality.

COLONIZATION OF MARS vs. OTHER POSSIBLE SOLUTIONS

In terms of possible solutions to the challenges facing humanity, Mars colonization is often compared to other options such as sustainable life on Earth, the use of renewable energy sources, and the exploration and exploitation of solar system resources. Although Mars colonization has captured the imagination of many, it is important to keep in mind the potential drawbacks and limitations of this option. Sustainable living on Earth, for example, could offer a more practical and realistic approach to addressing overpopulation and resource depletion. By adopting sustainable practices, such as energy-efficient technologies and waste reduction, we could reduce pressure on Earth's resources and ecosystem. Similarly, exploring and exploiting the resources of the solar system may offer a more feasible way to obtain vital resources such as minerals and water without the need to colonize another planet. Renewable energy sources, such as solar and wind, offer a promising solution to the problem of global warming without the need for large-scale colonization that may not be possible for many decades, if ever. While Mars colonization offers an exciting avenue for exploration and expansion beyond Earth, it should be considered alongside these other options as a complementary rather than exclusive solution.

In recent years, the debate surrounding the colonization of Mars has become increasingly popular as a possible solution to the various problems plaguing Earth. With population growth, depletion of natural resources and global warming, humanity is desperately seeking solutions to ensure its survival. In many ways, Mars represents the ideal candidate for such a colonization venture. Compared to other planets in our solar system, Mars is the closest planet to Earth that has a similar composition, with rocky surfaces and an atmosphere present. Its similarities to Earth are believed to make it more conducive to harboring life forms, as it has enough sunlight and heat to support plant life, and the planet's atmosphere is composed of approximately 95% carbon dioxide and 3% nitrogen, which can be manipulated to make it conducive to human life. Successful colonization of Mars is far from guaranteed, as many challenges remain. Mars is a harsh and inhospitable environment that requires careful planning, extensive and continuous research, innovation and adequate resources to ensure that life can be sustained in the long term. With all this in mind, the question remains whether the race to colonize Mars is worth the effort and resources, given the challenges ahead.

VII. TECHNICAL CHALLENGES

The technical challenges of Mars colonization are immense, as it will require the development of new technologies and the adaptation of existing ones. First and foremost is transporting people and equipment to Mars safely and efficiently, which currently involves a six-month journey. This will require the development of new propulsion systems that can provide enough thrust to accelerate spacecraft to high speeds and overcome the gravitational pull of the sun and other celestial bodies. Spacecraft will also have to be equipped with robust life support systems that can sustain human life throughout the journey and beyond.

Once on Mars, the colonists will have to be protected from a harsh and unpredictable environment. The Martian atmosphere is thin, composed mainly of carbon dioxide, and does not provide adequate protection against solar radiation or the extreme temperature variations that can occur. Structures will have to be designed to withstand high winds, dust storms and temperature extremes ranging from minus 100 degrees Celsius at night to minus 20 degrees Celsius during the day. Another key challenge is the production of food, water and oxygen on Mars. The soil on Mars is not suitable for growing crops and the planet's icy water is not easily accessible. Therefore, colonists will have to rely on advanced hydroponic and aeroponic technologies to produce sustainable food on the planet. Similarly, water will have to be extracted from subway sources or from the Martian atmosphere, which has only 1% of the density of Earth's atmosphere. A means of generating oxygen, either by electrolysis of water or by other means, will be needed to provide the colonists with the air they need to breathe. Communication with Earth will be essential for the survival and mental health of the colonists. Establishing a reliable and stable communication system between Mars and Earth will require the development of powerful radio communication systems and an extensive network of communication satellites. The great distance between the planets also means that any communication will be affected by a significant delay, which could add several minutes to the round-trip time, posing potential problems for emergency situations. Maintaining human health will be a major challenge, as the human body is not prepared to function in a low-gravity environment for long periods of time. The lack of gravity on Mars will cause a number of health problems, such as vision problems, muscle atrophy and bone loss, which could make even simple tasks difficult. Specialized medications and medical equipment will be needed to monitor vitals, diagnose and treat illnesses and injuries that may occur on Mars. The challenges will be even greater when the colonists begin to reproduce, as the implications of raising children in such a unique environment are undefined. Colonizing Mars is a daunting undertaking that poses numerous technical challenges. The journey will require new propulsion systems, advanced life support technologies and robust communication systems.

Structures capable of withstanding adverse weather conditions and extreme temperatures will have to be developed. To sustain life on Mars, effective food, water and oxygen production techniques will have to be developed, as well as medical equipment and drugs that can maintain human health in a low-gravity environment. Until we can address these technical challenges, Mars colonization cannot become a reality and we are therefore at a dead end. The continued study of Mars, therefore, is crucial, as it will provide invaluable insight and serve as the basis for the development of new technologies that will one day enable the colonization of Mars and the development of humanity beyond Earth.

TECHNICAL CHALLENGES: TRANSPORTATION, COMMUNICATIONS AND INFRASTRUCTURE

One of the main challenges associated with Mars colonization is the transportation of materials and people from Earth to Mars. The distance between Earth and Mars is enormous and it takes several months to travel between the two planets. This means that transportation methods must be developed that allow safe and efficient travel over long distances. Current methods of space travel, such as rockets, are not ideal for this task, as they are expensive and inefficient. The development of new technologies and transportation methods, such as space elevators or ion propulsion engines, will be essential for the successful colonization of Mars. Another major challenge associated with Mars colonization is communication. Due to the distance between Earth and Mars, communication delays can be as long as twenty minutes each way. This makes real-time communication between Earth and Mars impossible, creating unique challenges for the management and control of colony activities. To alleviate this problem, new communication technologies have been developed, such as the Mars Relay Network. This system enables reliable and consistent communication between Mars and Earth, making it possible to remotely operate equipment and conduct experiments on the Martian surface. In addition to transportation and communication problems, infrastructure is also a major obstacle to Mars colonization. The Martian environment is incredibly hostile, and the atmosphere is much thinner than that of Earth. This means that Mars colonization will require significant infrastructure development, including the construction of habitable structures, power systems, waste management facilities, and water and air management systems. These systems must be highly efficient and self-sustaining, as the colony will need to rely on its resources to survive. Developing a sustainable infrastructure will pose particular challenges to the Mars colony due to the planet's resource limitations. For example, Mars has limited water resources, and much of the water is frozen in its poles and soils. Therefore, the colony will have to develop efficient water management systems that can collect, treat and recycle as much water as possible. The use of renewable energy sources, such as solar and wind, will also be crucial to the colony's success, as these sources can provide the energy needed for most of the colony's systems. The Martian atmosphere is composed mainly of carbon dioxide, with no breathable gases. This means that the colony will need to develop air management systems that can produce breathable air for the inhabitants. This can be achieved through the use of plants that can produce oxygen through photosynthesis, or through the use of mechanical scrubbers that can remove carbon dioxide and other harmful gases from the atmosphere. The colonization of Mars also poses significant challenges to the physical and mental health of the colonists. Due to the planet's lower gravity and less dense atmosphere, colonists will be exposed to higher levels of radiation than on Earth. This increased radiation exposure can lead to higher rates of cancer and other radiation-related health problems. Colonists must be protected from these risks by developing new shielding technologies or by using subway or shielded habitats. Living in an isolated, confined and extreme environment can cause significant mental health problems for colonists. The psychological stress of living in a harsh and dangerous environment, separated from friends and family on Earth, can lead to depression, anxiety, and other mental health problems. Developing effective mental health strategies and support systems to cope with the unique challenges of life on Mars will be essential to the colony's success. The technical challenges associated with Mars colonization are numerous and complex, from transportation and communications to infrastructure and the physical and mental health of the colonists. The development of new sustainable technologies and systems will be crucial to the success of the Martian colony, and lessons learned from Mars colonization will likely have important applications for sustainable development on Earth. Scientists, engineers and policy makers working together to overcome these challenges will pave the way for a new era of human exploration and discovery, ensuring the survival and prosperity of humankind for centuries to come.

ASSESSMENT OF THE TECHNOLOGICAL ADVANCES NEEDED TO SUCCESSFULLY COLONIZE MARS

The colonization of Mars is a colossal undertaking that requires a range of technological advances, some of which already exist and some of which are under development. Some of the main areas of interest are the construction of sustainable and self-sufficient habitats, transportation and logistics, energy generation and storage, and food production. First, habitats must have some type of artificial life-support system that mimics the terrestrial ecosystem. This involves the use of technologies such as air conditioning and filtration, water purification systems, waste management and other processes essentials for life. Habitat construction will also have to take into account the harsh Martian living conditions, which include surface radiation, limited resources, corrosive dust, and extreme hot and cold temperatures. Therefore, highly innovative materials that can withstand physical, chemical and environmental stresses will be needed. Second, transportation and logistics will be crucial to colonize Mars. Given that the planet is 54.6 million kilometers away, it would take months for spacecraft to reach Mars, and the issue is further complicated by the timing of launch windows and atmospheric conditions. Future Mars missions could reduce the wait time with rapid transit designs, such as nuclear-powered spacecraft, but suitable landing sites, surface vehicles, and material handling equipment will still be needed to reach and exploit Mars' surface resources. Spacecraft that support long-duration spaceflight and cargo delivery, such as reusable space launch vehicles, will also be needed. These vehicles will need to provide ample support for both humans and cargo, adequate power supply, and communication that can withstand the long and harsh transportation environment. Third, effective and efficient power generation and storage systems are essential. It is paramount to meet the energy needs of Martian colonization facilities, cargo transport, surface mining and refining operations, space hardware manufacturing for Mars operations, and all life support systems. Multiple options for power generation include solar panels, nuclear reactors, and wind turbines, among others, depending on the scope and needs of the mission. Energy storage systems, such as batteries or fuel cells, are critical, especially given the dynamic power demands during the launch and landing phases of a Mars mission. Fourth, the ability to produce food sustainably will be crucial to ensure a prolonged stay on the planet. Growing crops on Mars is challenging due to the harsh and unfavorable surface conditions. Innovative solutions such as the creation of hydroponic and aeroponic facilities and other controlled environment systems could prove effective. With these systems, plants can be grown in a soilless medium optimized for growing conditions even in a Martian environment. The systems should also be able to regulate light and temperature to simulate Earth-like conditions and avoid plant stress caused by Mars' limited resources and low atmospheric pressure. Animals, such as fish, could also be grown in closed aquatic environments that can further supplement the colonists' nutritional needs. In summary, the colonization of Mars is a complex process that requires innovative technological advances in several areas such as habitat construction, transportation and logistics, energy generation and food production. These advances will not only have to be reliable and effective, but also sustainable in the long term, as conditions on the red planet differ greatly from those on Earth. Technological advances will also need to be able to support and sustain multiple industries, from research, manufacturing and mining to tourism and even residential living. With the right technologies and resources, humanity can achieve a path to Mars colonization that has the potential to ensure long-term survival and expansion beyond planet Earth.

TECHNICAL CHALLENGES OF COLONIZING MARS vs. OTHER POSSIBLE SOLUTIONS

Mars colonization has been suggested as a solution to humanity's urgent need to find an alternative habitat. Although it promises great potential, the technical challenges involved are enormous, and comparison with other possible solutions reveals both the advantages and limitations of Mars colonization. One of the most significant challenges of Mars colonization is its inhospitable environment. Mars has a thin atmosphere of mostly carbon dioxide, no liquid water, and temperatures ranging from -140 to 20°C, which poses a significant challenge to human survival. Although some scientists believe that terraforming Mars, which involves altering its surface environment to make it more Earth-like, could offer a solution, this is currently beyond our technological capabilities. Establishing a colony on Mars would require the development of advanced technologies, as well as a reliable means of transporting people, supplies, and equipment.

In comparison, other possible solutions, such as building underwater cities, present a less extreme environment. Although building underwater cities involves some technical challenges, such as designing structures that can withstand deep water pressure, the environment is more accommodating than Mars. For example, the ecosystem is more reliable, with fish and other aquatic life providing a potential food source. Water pressure could provide a protective barrier against cosmic radiation. Similarly, building floating cities in the oceans, which has been suggested as a possible solution, involves fewer technical challenges than colonizing Mars. The environment is more hospitable to human life and many of the technologies needed to make floating cities possible, such as solar and wind power, are already available. Floating cities could also be mobile, meaning they could be moved to avoid natural disasters such as hurricanes or rising sea levels. Another possible solution is space habitats, which propose the construction of structures that would allow humans to live and work in space. Space habitats are more technically feasible than colonizing Mars and could provide a home for humans in space, thus avoiding some of the risks associated with long-duration spaceflight. These habitats could be built with existing materials and technologies and would not require the massive infrastructure and terraforming that would be involved in Mars colonization. Space habitats could also offer a solution to the problem of resource depletion, as they could provide a means of harvesting and utilizing space resources. The colonization of space would allow scientists to conduct research that is not feasible on Earth, such as observations of distant stars and galaxies, which could lead to breakthroughs in science and technology. It should be noted that each of these solutions comes with its own set of challenges and limitations. Building underwater or floating cities would require considerable investment, and there are still questions about the long-term sustainability of these habitats. Similarly, space habitats could pose health risks associated with long-term exposure to zero-gravity environments or radiation. Space habitats do not address the issue of resource depletion on Earth, which could continue to worsen if viable solutions are not implemented. Overall, while Mars colonization presents a promising solution to humanity's need for a new habitat, it is not without its challenges. In comparison, other potential solutions such as underwater or floating cities, as well as space habitats, have less formidable technical challenges to overcome. Each solution has its own limitations and would require considerable investment, research and development. It is therefore essential that all potential solutions are carefully considered, with a focus on finding viable, sustainable and long-term solutions to the pressing problems facing humanity.

The depletion of the Earth's natural resources, coupled with overpopulation and global warming, has prompted the search for solutions to ensure humanity's survival. For many, the colonization of Mars a viable option. The idea is that, if we can establish a human settlement on Mars, we could reduce pressure on Earth's resources and provide a backup plan in case of global catastrophe. The prospect of colonizing Mars is not without its difficulties. Even with technological advances, the journey to Mars is dangerous and fraught with obstacles that could endanger the lives of potential colonists. The long-term sustainability of a Mars settlement remains uncertain, as it is unclear whether humans can adapt to living in such a harsh and unfamiliar environment. Despite these challenges, the prospect of colonizing Mars remains a beacon of hope for many who see it as a way to preserve our species and advance scientific knowledge. To achieve this goal, we will need to continue to invest in research, technology, and infrastructure, while carefully considering the ethical implications of expanding our influence on a new world. The colonization of Mars represents a bold and necessary step forward for humanity, which could serve as a means to ensure our survival and advance our understanding of the universe.

VIII. SOCIAL AND ETHICAL IMPLICATIONS

The ethical and social implications of Mars colonization are immense and far-reaching and must be carefully considered along with the practical and technological obstacles to successful human settlement on the Red Planet. Of particular concern is the fact that colonization of Mars would almost certainly involve exploitation of the planet's natural resources, which could result in a new form of colonialism and exploitation against an alien world. This raises profound questions about humanity's responsibility towards other life forms and the environment, and about the ethical limits of our technological and economic aspirations. Another ethical consideration is the impact of Mars colonization on the social fabric of humanity. Assuming successful colonization of Mars, it is likely that those who can afford to migrate there will do so, leaving behind those who cannot or choose not to. This could exacerbate inequalities between the world's "haves" and "have-nots" and lead to political and social conflict. Colonization of Mars may require a strong centralized government to ensure the survival of its inhabitants, resulting in a new form of authoritarianism or even dictatorship. The social and cultural implications of Mars colonization are equally complex. The creation of a new human society on a distant planet would almost certainly involve the abandonment of traditional cultural norms and social conventions, leading to the creation of new hybrid cultures that would be difficult to foresee or conceive of in advance. This could lead to a loss of cultural diversity and richness, as well as the erosion of traditional social values. At the same time, the mere fact of colonizing Mars would be a profound cultural and social achievement, which could inspire humanity to achieve great things together, in the name of progress and survival. The question of whether or not colonizing Mars is ethical comes down to our responsibility as a species. As we face the challenges of overpopulation, climate change and resource depletion, we are being asked to balance our aspirations for progress with our responsibility to the planet and all its inhabitants. While Mars colonization may seem like the ultimate solution to our problems, it is not without significant ethical and social implications that must be considered and addressed. Only by recognizing and addressing these implications can we move toward a future in which we can live in harmony with both Earth and the universe at large.

SOCIAL AND ETHICAL IMPLICATIONS: IDENTITY AND CULTURAL HERITAGE OF HUMANITY

The decision to colonize Mars could have a significant impact on humanity's identity and cultural heritage. Leaving Earth behind to start anew on another planet raises questions about what it means to be human and our place in the universe. Colonization of another planet could fundamentally change our understanding of ourselves and our relationship to the natural world. It could lead to the loss of cultural heritage as the focus shifts to the creation of a new society on Mars. One of the major social and ethical implications of Mars colonization is the potential impact it may have on humanity's identity. Leaving Earth behind and starting over on Mars could result in a redefinition of what it means to be human. Living in a completely different environment, with different resources and challenges, could force us to adapt and evolve in ways we cannot yet imagine. Some argue that this could lead to a loss of our connection to the natural world and a shift toward a more artificial and mechanical existence. Others argue that it could lead to a new appreciation of our place in the universe and a revitalization of our sense of wonder and curiosity. In addition to the impact on our identity as humans, the colonization of Mars also raises questions about the preservation of our cultural heritage. As we focus on building a new society on Mars, we may neglect or even lose aspects of our cultural heritage that have defined us for generations. The physical distance between Earth and Mars, as well as the time and resources required to establish a new colony, could

99

exacerbate this problem. Even if efforts are made to preserve our cultural heritage, for example, through digitization and information storage, it is unclear to what extent they can be effective in the long term. The potential loss of cultural heritage is not a new concept and has been a concern throughout human history. The unique circumstances of Mars colonization present a number of challenges and opportunities that require careful consideration. In this regard, we must be mindful of how we approach Mars colonization and ensure that our actions are not detrimental to our cultural heritage. Beyond these immediate concerns, the long-term impact of Mars colonization on humanity's identity and cultural heritage is difficult to predict. It is possible that our understanding of ourselves and our place in the universe will evolve in beneficial ways, leading to a new era of scientific discovery and cultural exchange. Another possibility is that we will become more insular and disconnected from the natural world, leading to a loss of biodiversity and natural resources that will be detrimental to our survival. Whatever the outcome, the social and ethical implications of Mars colonization require deep reflection and ongoing dialogue. We must be aware of the potential benefits and risks, and ensure that our actions conform to our values and principles as a society. While the decision to colonize Mars may be motivated by the need for survival, it must also be underpinned by a commitment to social and environmental responsibility. The debate on the social and ethical implications of Mars colonization is complex and multifaceted. There are concerns about the impact on human identity and cultural heritage, as well as the potential loss of biodiversity and natural resources on both Earth and Mars. At the same time, there is hope that the colonization of Mars could lead to a new era of scientific discovery, innovation and cultural exchange. The decision to colonize Mars is a reflection of our collective values and priorities as a society, and should be approached with care and consideration for all parties involved.

IMPACT ON HUMAN RIGHTS, SOCIAL JUSTICE AND EQUALITY

The colonization of Mars is a complex and multifaceted undertaking that raises numerous ethical and social issues, including those related to human rights, social justice and equality. First, the colonization of Mars must balance the interests of all parties involved in the colonization process: the colonizers, indigenous life on Mars, and the human population on Earth. To ensure that the colonization process is carried out in an ethical manner, safeguards must be put in place to manage the risks associated with potential exploitation of the Martian environment and its resources. It is important to remember that the idea of colonizing another planet raises the question of who decides which lives and life forms are considered valuable and which lives and life forms are disposable or expendable. This highlights the larger ethical questions surrounding the expansion of human presence in space, and how the associated technologies, policies, and decisions may privilege some populations over others. The colonization of Mars raises several issues related to social justice. For example, the selection of the first Martian colonists may be based on criteria that institutionalize inequality and benefit the wealthy or powerful. This may deepen existing disparities in access to resources and opportunities. Colonization may also exacerbate gender and racial inequalities. It is no secret that the space industry remains predominantly white and male, and exploiting Martian resources for profit may further disenfranchise marginalized communities. In turn, this could lead to social and political instability, as has happened in the past. Another important consideration is the impact that colonization may have on human rights. Although the UN Declaration of Human Rights does not apply directly to the Martian environment, the principles underlying this document, such as recognition of human dignity, freedom, equality, and justice, should continue to guide human activities on Mars. Ensuring respect for human rights during colonization will be particularly complex because of the institutional and technological constraints that will exist on the first colonies. For example, the availability of resources such as water, food, and shelter may be limited in the early phases of colonization, which could lead to some individuals occupying positions of power to exert control over scarce resources. In the worst case, this could lead to human rights violations such as exploitation, discrimination and even slavery. There is the potential question of how colonization of Mars may affect those who remain on Earth. If priorities continue to be skewed towards the ambitions of the rich and powerful, it could contribute to further suffering for those who suffer from the effects of climate change, reduced access to resources, and political instability. This is especially relevant when thinking about how interplanetary colonization efforts will affect populations that are already experiencing the detrimental effects of unsustainable practices on Earth and whose displacement may make them more vulnerable. It is clear that Mars colonization is a complex and multifaceted undertaking that raises numerous social, ethical, and human rights issues. It is important to weigh the benefits of interplanetary exploration and colonization and take them into account in design, planning and policy development. The best way to ensure that colonization is carried out ethically and fairly is to involve a broad base of stakeholders in the decision-making process across sectors (e.g., government, private sector, civil society). Human efforts must be mindful of the inevitable trade-offs that will arise and continue to aspire to outcomes that prioritize fairness, equity and ethical considerations over immediate gains and power.

SOCIAL AND ETHICAL IMPLICATIONS vs. OTHER POSSIBLE SOLUTIONS

While Mars colonization may seem like a promising solution to our current problems, it is important to consider the potential social and ethical implications of this approach. First, Mars colonization would require significant financial investment and raises the question of whether governments should prioritize it over other pressing issues such as poverty reduction, healthcare and education. Mars colonization could entail resource exploitation and the possible displacement of native Martian species. This raises ethical issues about the appropriate use of resources and the consequences of disturbing an ecosystem about which we know very little. In comparison, solutions such as reducing carbon emissions and investing in renewable energy sources are much more environmentally friendly and raise fewer ethical issues. Another possible solution is the exploration of alternative forms of energy, such as nuclear or fusion power. Although nuclear energy has its own ethical implications, such as the risks associated with nuclear waste, it is generally considered a more viable solution than colonizing Mars, as it has already been successfully implemented in many countries around the world. Fusion energy, which consists of harnessing the energy of nuclear reactions in a controlled manner, is still in the development phase, but has the potential to be a game changer in terms of energy production. Unlike the colonization of Mars, these solutions have a much lower risk of environmental damage and habitat displacement. The colonization of Mars also has social

107

implications. The idea of living on another planet may sound exciting, but it is important to consider who will have the opportunity to do so. As with most technological advances, the wealthy would likely be the first to benefit from Mars colonization, leaving the rest of society behind. This raises concerns about social inequality and the possibility of a widening gap between rich and poor. In contrast, solutions such as sustainable agriculture, which involve empowering local communities to produce their own food and reduce dependence on large corporations, have the potential to promote a more equitable distribution of resources. It is important to consider the feasibility of colonizing Mars in the first place. Although technological advances have brought us closer to making this a reality, there are still many unknowns when it comes to living on another planet. Factors such as the impact of long-term space travel on the human body, the ability to produce sustainable food and water sources, and the ability to adapt to a completely different climate and environment raise questions about the feasibility of colonizing Mars. In comparison, solutions such as carbon capture and storage, which involves capturing carbon dioxide from power plants and storing it underground, have already been tested and successfully applied in certain areas. Overall, while colonizing Mars may seem like an exciting prospect, it is important to consider the potential social and ethical implications, as well as the feasibility of this approach. While solutions such as renewable energy, nuclear power, sustainable agriculture, and carbon capture and storage may not be as glamorous, they have the potential to reduce our impact on the environment and promote a more equitable distribution of resources. As we continue to search for solutions to our current problems, it is important to critically evaluate all potential approaches and consider the long-term consequences they may have for our planet and its inhabitants. There is no doubt that the Earth's ever-increasing population, combined with the depletion of natural resource sources and the global threat of climate change, has posed existential crises that threaten the very survival of humankind. Despite numerous efforts to mitigate these challenges, the reality appears to be that the Earth may not be able to sustain human life for much longer. From this perspective, hope for humanity's survival has shifted to the exploration and colonization of Mars. Although Mars has its own challenges, it remains a viable alternative for human survival. The colonization project poses unique opportunities in terms of scientific exploration, technological advancement, economic development and, most importantly, ensures the long-term survival of humanity beyond Earth. This prospect has attracted the attention of researchers, engineers and enthusiasts around the world, culminating in various initiatives aimed at making Mars colonization a reality. The success of the project depends on several interdependent factors such as transportation, infrastructure and resource management, all aimed at achieving a sustainable and viable human habitat on Mars.

One of the main concerns for Mars colonization is transportation. Mars is approximately 140 million miles from Earth, which translates to about six months of travel. There is currently no known technology that can transport humans to Mars in less time, but scientists are researching and testing possible ways to reduce travel time. The goal is to make transportation to Mars as fast and efficient as possible. With the advancement of space exploration technology, researchers believe that within a few decades there could be a technology to replace or complement current spacecraft technology to improve propulsion and reduce travel time and make such a project practical. For example, ion propulsion is a promising technology that is touted as a replacement for traditional rocket propulsion systems. Ion propulsion facilitates smoother and faster travel through space, but its application in larger spacecraft capable of carrying humans is still in the research and testing phase. The success of the Mars project, therefore, would require significant investment in research and innovative technology that can facilitate fast, efficient and safe transportation to Mars. Another critical factor to consider is the development of infrastructure on Mars. A human habitat on Mars would require the construction of major infrastructures, from power generation, water and air supply, food production and waste management systems. These infrastructures would require constant maintenance and upgrading, given Mars' hostile environment. Mars lacks a protective magnetosphere, making it particularly vulnerable to solar radiation and other harmful cosmic rays. This radiation poses a high risk of cancer, DNA mutation, and other health hazards to the inhabitants of Mars. Therefore, a critical aspect of infrastructure development would be to establish appropriate radiation shields or other technologies to reduce exposure to cosmic radiation and maintain astronaut health and safety. Establishing a sustainable and reliable infrastructure on Mars would require robust technology and innovative engineering solutions that sustainably harness Mars' resources and manage them efficiently. The challenge of resource management is critical to establishing a viable and sustainable human habitat on Mars. Despite its similarities to Earth, Mars poses unique challenges in terms of resource access and management. For example, water, a critical resource necessary for human survival, is scarce on Mars. To provide water and other necessary resources to human inhabitants, the Martian environment would require significant resource management solutions including recycling and regeneration of resources such as water and efficient use of other resources such as minerals and metals. A food production infrastructure would have to be created to ensure an adequate supply of fresh food for the human inhabitants of Mars. Advances in engineering and agricultural technology are important for studying crop growth in the harsh Martian environment. Successful resource management depends on the availability of adequate resources, the development of innovative technologies, and socioeconomic arrangements that ensure equitable and efficient distribution and management of resources on Mars. The exploration and eventual colonization of Mars offers a glimmer of hope for humanity facing monumental challenges that threaten its very existence. The colonization project holds immense promise in several respects, ranging from scientific exploration, economic development and technological advancement to the long-term survival of humanity beyond Earth. The success of the project depends on numerous interdependent factors that require significant investment in research, innovative technology, and comprehensive infrastructure development and resource management. The opportunity presented by the Mars colonization project is enormous, and the potential benefits far outweigh the challenges ahead. With collective effort, investment in innovative research and development, and determination, humanity can overcome Earth's challenges and establish a sustainable and viable human habitat on the Red Planet.

IX. THE ROLE OF PRIVATE AND PUBLIC ENTITIES

Faced with the daunting challenge of colonizing Mars, the role of both private and public entities cannot be overemphasized. While the public sector plays a vital role in funding and exploring the Red Planet, the private sector can facilitate the commercial viability of the Mars mission. Public entities such as NASA, the European Space Agency and the China National Space Administration are the major players in Mars exploration, with NASA leading the way. Since the 1960s, NASA has explored Mars, with several probes and rovers landing on the planet's surface. The agency has numerous Mars missions underway, such as the Mars 2020 Perseverance rover mission and the Mars Sample Return mission, among others, aimed at collecting scientific data to support future manned missions to Mars. On the other hand, the private sector sees the Mars mission as an opportunity to commercialize the space industry. Entrepreneurs such as Elon Musk, Jeff Bezos and Richard Branson have invested huge sums of money in Mars colonization, which could unlock economic opportunities in space tourism, mining and research. Private entities such as SpaceX, Blue Origin and Virgin Galactic play different roles in the colonization of Mars. SpaceX, founded by Elon Musk, has focused on Mars colonization, with a mission to make humanity a multi-planetary species and reduce the risks of extinction on Earth. SpaceX has been instrumental in developing the technology and infrastructure needed for Mars colonization, such as the Starship spacecraft, the Super Heavy rocket and the

Mars Base Alpha concept. Musk's plan is to make the Mars mission affordable for humans, which involves developing reusable spacecraft and rockets. Similarly, Blue Origin, founded by Amazon CEO Jeff Bezos, is interested in exploring Mars, the Moon and asteroids, with a view to achieving sustainable manned spaceflight. Blue Origin has developed reusable rockets and spacecraft, and is working on lunar lander technology in collaboration with NASA's Artemis program. Virgin Galactic, meanwhile, has focused on space tourism, giving people the opportunity to try spaceflight at a premium price and thereby generating revenue for space exploration and research. Private sector involvement is not only limited to technology and infrastructure, but also extends to funding and research. Private entities are funding their own Mars missions, collaborating with government agencies and universities to advance Mars research. For example, SpaceX has been working with NASA and other private entities such as Boeing, Lockheed Martin and Dynetics in developing different aspects of the Artemis program. Blue Origin, through its Blue Moon mission, aims to land on the Moon and potentially use it as a gateway to Mars. Private entities are also helping to reduce the risk of the Mars mission, as they are willing to take on high-risk projects that governments might shy away from. This, in turn, can pave the way for other entities to invest in Mars exploration and colonization, ultimately making it more feasible. The public sector plays an important role in funding and regulation. Government allocation of funds to space exploration is essential, as the cost of the Mars mission is in the billions of dollars. NASA's budget for the Mars mission was $2.8 billion in 2021, a significant portion of its overall budget of $23.3 billion. This funding goes toward spacecraft development, research, and paying the salaries of scientists, engineers, and other personnel. Government investment in Mars exploration provides an avenue for scientific advancement that would not be possible without public funding, as the benefits of investment can be intangible and long-term. The government also regulates space activities to ensure international cooperation, collaboration, and safety. The United Nations Outer Space Treaty, ratified by 110 countries, regulates and guides the activities of states in the exploration and use of space, ensuring the peaceful and safe use of outer space and celestial bodies. Governments, therefore, have an important role to play in regulating and maintaining international cooperation in the mission to Mars. Another key role of public entities in the Mars mission is diplomacy. The Mars mission is a global mission, with several countries playing an important role. Countries such as the United States, Russia, China, the European Union, and India have developed their Mars missions and their research and technology capabilities. Cooperation in space exploration can promote diplomatic relations and forge partnerships that extend to other industries in the global economy. Public diplomacy can thus facilitate international collaborations that advance the Mars mission and promote economic development. The Mars mission requires the collaboration of public and private entities, each playing a different role. The role of government in funding, regulation and diplomacy and the role of the private sector in developing technology, infrastructure and funding are critical to the success of the Mars mission. By combining resources, technological expertise and capital investment, the goal of colonizing Mars can be achieved. This can unlock new economic opportunities, advance scientific knowledge and ensure humanity's survival in the face of global challenges.

PRIVATE AND PUBLIC ENTITIES

In the quest to colonize Mars, both private and public entities play crucial roles. Private entities, such as SpaceX and Blue Origin, have pioneered a new era of space exploration, focused on opening up commercial space travel and reducing the cost of launching missions. In contrast, public entities, such as NASA and the European Space Agency (ESA), are focused on scientific research and exploration of the universe. This dichotomy gives rise to an interesting dynamic, as private companies compete to offer the most cost-effective solutions to reach Mars, while public entities use their budgets to ensure that scientific exploration of the planet continues. Private entities such as SpaceX, with their reusable rockets, are driving down the cost of space travel and making Mars colonization a more feasible endeavor. At the same time, public entities such as NASA and ESA are seeking to develop the technology necessary for long-term sustainability on Mars, including the development of habitats, transportation systems and power generation sources. Both private and public entities must work together to make Mars colonization a reality. Private companies can provide the ambition and funding needed to launch space missions, while public entities can provide the scientific expertise and oversight needed to ensure the success of the colonization process. Together, they can create a sustainable Martian colony to serve as a potential future home for humanity. Collaboration between the public and private sectors on a Mars colonization project could lead to new technological breakthroughs and innovations that could be used in future space exploration missions. Humanity's survival may depend on our ability to successfully colonize Mars, and both private and public entities have a critical role to play in realizing that vision.

ASSESSMENT OF CONFLICTS BETWEEN PRIVATE AND PUBLIC ENTITIES

It is crucial to understanding the challenges of space exploration. The space industry is growing rapidly, led by private entities such as SpaceX, Blue Origin and Virgin Galactic. As private companies begin to spearhead space exploration, conflicts regarding property rights, national security and even ownership of extraterrestrial resources could arise between government agencies and private entities. The United Nations Outer Space Treaty, which regulates the exploration and use of outer space, recognizes that space is the common heritage of all mankind. Private companies may attempt to claim and exploit resources found on other planets. This could lead to disputes with governments over ownership and exploitation rights, as well as with other private companies. Private entities also pose national security concerns, especially for countries with significant geopolitical interests in space. In the event of conflict between these countries, private entities could become a liability, as they may not be bound by national security interests or diplomatic boundaries. The issue of intellectual property rights in space is complicated. Private entities are likely to invest significant resources in the development of new technologies for space exploration and resource exploitation. These entities should not infringe on government patents, copyrights or other proprietary technologies that may exist. Similarly, governments must also protect the intellectual property rights of private entities in order to incentivize greater investment in space exploration. Overall, balancing the interests of private entities and governments in space exploration is crucial to the continued growth of the industry and to ensuring that humanity can continue to explore and use the resources of outer space with confidence.

PRIVATE vs. PUBLIC ENTITIES

As the prospect of human colonization of Mars becomes increasingly plausible, there are two key categories of institutions vying for a role in the project: public agencies such as NASA and private companies such as SpaceX. While both public and private entities have come a long way in developing the technology and expertise needed for Mars exploration, there are several key differences between their approaches that are worth examining.

One of the most significant contrasts between public and private entities is the way they are funded. Public organizations such as NASA are funded by governments, which means they are subject to the whims of politicians and the availability of public funds. In contrast, private companies are funded by investors, which allows them a greater degree of flexibility in terms of their operations. This means that private companies can often act more quickly and efficiently than public entities, as they do not need to go through the same bureaucratic processes. SpaceX has been able to make rapid progress in its efforts to colonize Mars thanks in part to the abundant funding it has received from private investors. Although NASA has long been interested in Mars exploration, its budget has been subject to significant fluctuations over the years. This has made it difficult for the agency to establish a consistent course of action, as each presidential administration has had different priorities for NASA. In contrast, SpaceX has a clear vision for its Mars colonization efforts, and has been able to invest heavily in research and development without any of the political uncertainties that can hamper the long-term plans of a public entity. Another key difference between public and private entities is their approach to collaboration. Public organizations such as NASA often rely on collaboration with other organizations, including other government entities and private companies. This can make it difficult for public entities to maintain consistency and control over their projects as they work with a wide range of stakeholders. In contrast, private companies such as SpaceX have a great deal of control over their operations, allowing them to work more independently and make decisions quickly. This can be especially useful when it comes to research and development, as a company like SpaceX can focus on developing the technologies it needs without being constrained by the demands of external stakeholders. At the same time, public entities such as NASA have access to a wider range of resources than private companies. For example, NASA has been able to tap into the experience and expertise of generations of astronauts and scientists, as well as the resources of the U.S. government. This has enabled the agency to develop technologies and strategies based on decades of experience and research. In contrast, private companies such as SpaceX are starting from scratch when it comes to Mars colonization, and must rely on their own research and development efforts to move forward. Perhaps the biggest advantage that private companies have over public entities when it comes to Mars colonization is their ability to take risks. Public entities such as NASA are under a great deal of scrutiny, both from politicians and the general public. This can make it difficult for them to take risks with unproven technologies or strategies, as they may be subject to criticism if things go wrong. Private companies, on the other hand, are free to take risks in their efforts to colonize Mars without fear of the same level of criticism. This can be especially useful when it comes to developing new technologies or strategies, as companies are free to experiment and innovate without worrying about the potential consequences of failure. Both public and private entities have made significant progress in their efforts to colonize Mars. While public entities such as NASA have traditionally led the way in space exploitation, the rise of private companies such as SpaceX has opened up new possibilities for Mars colonization. The ideal approach to Mars colonization may lie somewhere between these two poles, leveraging the strengths of public and private entities to create a comprehensive and coordinated effort that can achieve the goal of human settlement on Mars. The concept of colonizing Mars has been a topic of debate in the scientific community for decades, sparked by the idea of a backup plan for humanity in the event of catastrophes on Earth. As the problems of overpopulation, natural resource depletion and global warming continue to be of concern, the idea of colonizing Mars is increasingly seen as a possible solution to these problems. With companies such as SpaceX and NASA devoting resources to explore the possibility of establishing a colony on the Red Planet, hope is growing that humanity can ensure its survival beyond Earth. The benefits of a Mars colony could include developing new technologies, conducting scientific research, and the possibility of sustainable life in a hostile environment. There are also numerous challenges, such as the physical and psychological toll of long-term space travel and the difficulties of building and maintaining a self-sustaining colony on a barren planet with a hostile environment. But with the future of humanity at stake, Mars exploration offers a glimmer of hope for a better tomorrow.

X. THE COSTS OF MARS COLONIZATION

One of the biggest challenges of Mars colonization is the high economic cost involved. It is estimated that establishing a sustainable human settlement on Mars would cost billions of dollars. NASA's current plan to send humans to Mars in the 2030s is estimated to cost at least $20 billion. This is only the initial cost of sending a manned mission to Mars and does not include the cost of building a permanent settlement on the red planet. The cost of building the infrastructure necessary for human life on Mars, including human life on Mars, including habitats, life support systems and transportation. The colonization of Mars also entails operational costs. Once a permanent settlement is established, it will need a constant supply of resources and regular maintenance to ensure its sustainability. This will require a significant investment of resources and manpower. The cost of transporting people and supplies to and from Mars will be extremely high. The distance between Earth and Mars means that a round-trip mission to Mars will take between 1 and 3 years, depending on the position of the planets in their respective orbits. The cost of fuel would be enormous. In addition to the financial costs, colonization of Mars poses significant technological challenges. Establishing a self-sufficient human settlement on another planet would require a range of advanced technologies that do not currently exist. One of the biggest challenges is to develop a sustainable life support system that can provide the necessary resources, such as air, water and food, to sustain human life on Mars. This will require a range of innovative technologies, such as advanced hydroponic systems to grow crops in the harsh Martian environment, and sophisticated recycling systems to recover and reuse waste products. Developing and testing these technologies will likely require significant investments in research and development. Another technological challenge is the development of a transportation system that can safely and efficiently move people and supplies to and from Mars. This will require the development of advanced propulsion systems that can cope with the extreme distance and harsh conditions of space travel. Current space exploration technologies, such as chemical rocket engines, are not powerful enough to make the journey to Mars efficiently. The development of more advanced propulsion technologies, such as nuclear rockets or ion propellants, will require significant research and development and testing in space. Despite the significant challenges posed by the colonization of Mars, many argue that it is a necessary step for humanity to ensure its long-term survival. Overpopulation, depletion of natural resources and global warming threaten the existence of our species on Earth. Colonizing Mars could provide a backup plan for humanity's survival in the event of catastrophe, such as the impact of a large asteroid or a massive nuclear war. It could also offer new opportunities for scientific research and technological innovation. For example, the study of the unique geology and environment of Mars could provide information on the history and evolution of the solar system and the universe as a whole. Developing the technologies needed for Mars colonization could also have far-reaching benefits, such as more efficient and sustainable transportation systems, advanced life support systems, and new materials and manufacturing processes. Colonizing Mars is a daunting but necessary challenge for humanity. The financial costs associated with colonizing Mars are high, but are outweighed by the potential benefits of ensuring the long-term survival of our species and providing new opportunities for scientific research and technological innovation. Significant technological challenges must be overcome before a sustainable human settlement can be established on Mars. These challenges include the development of advanced life support systems, transportation systems, and sustainable resource management systems. Addressing these challenges will require significant investments in research and development, but the potential benefits justify the expense. The colonization of Mars represents a bold step forward for humanity and a possible solution to some of the greatest existential threats facing our species.

DEBATE ON THE COSTS OF COLONIZING MARS

The financial costs associated with Mars colonization are immense and cannot be underestimated. It is estimated that the initial phases of colonization alone will exceed billions of dollars. Currently, the cost of sending an unmanned mission to Mars is around $500 million, while sending a manned mission could cost as much as $2.5 billion. To establish a permanent human presence on the red planet and support the growth of a self-sustaining Martian civilization, the costs could exceed trillions of dollars. Significant funding will be needed for research and development of new technologies, such as propulsion systems, habitats, and life support systems, necessary to establish and maintain human life on a planet with hostile environments. Spacecraft construction and fuel costs will have to be taken into account, as well as provisions for sustaining human life on a planet with hostile environments. Spacecraft and fuel construction costs will also have to be taken into account, as well as provisions for the development of infrastructure on the planet's surface, such as power grids, communication and transportation networks. Problems such as resource scarcity, extreme weather conditions and the hazards associated with prolonged exposure to radiation and microgravity could arise on Mars. All of these factors make the financial investments required to establish a self-sustaining colony on Mars of a magnitude and urgency never before seen in human history. The journey to Mars is not only expensive, but also requires a considerable amount of effort and resources. The considerable distance between Earth and Mars means that astronauts will have to be in space for more than eight months before reaching the Martian surface. The long journey will require advanced technical systems and significant amounts of food, water and oxygen supplies. Although innovations in propulsion technology have enabled spacecraft to reach Mars much faster than previous missions, significant advances are needed to avoid the physical and mental damage associated with prolonged exposure to microgravity environments. Without advanced technological and medical advances, prolonged exposure could cause bone loss, muscle atrophy, and weakened immune systems in astronauts. The costs of a Mars mission could have a domino effect on public spending priorities. At present, public budgets are already stretched to capacity, with many countries facing debt and deficits. Some critics have argued that the funds needed for a Mars mission should be redirected to address pressing problems on Earth. For example, funds allocated to a single Mars mission could alleviate the plight of millions of people who lack access to basic human needs such as food, water and health care. Critics also argue that governments should prioritize solving pressing economic and political problems on Earth before investing in space exploration. Proponents argue that innovations developed during space exploration have led to improvements in several fields, such as medicine, telecommunications and environmental sustainability. They argue that a mission to Mars could lead to similar advances and benefits.

One possible way to reduce the costs of a Mars mission is international cooperation and public-private partnerships. The international community has already come together on other occasions to achieve common goals. Examples include the International Space Station, which several countries collaborated to build and maintain, and the Apollo program, which successfully landed humans on the Moon. Similar international collaboration could be key to the success and cost-effectiveness of a Mars mission. Cooperation between different cultures, political systems and financial frameworks can face significant obstacles.

Another possible way to fund Mars exploration is through public-private partnerships. Private companies such as SpaceX and Blue Origin, led by Elon Musk and Jeff Bezos, respectively, have already expressed strong interest in exploring Mars. With their vast resources and expertise, private companies could work with governments and international organizations to establish a self-sustaining colony on Mars. Partnerships could reduce the financial burden on each country while enabling significant investments in research, innovation and infrastructure. Private companies could also generate new sources of revenue by investing in commercial opportunities on Mars, such as mining rare minerals or establishing self-sustaining tourism industries. The financial costs associated with Mars colonization are substantial and require significant investments in research, innovation, transportation and infrastructure development. Governments, international organizations and private companies must work together to reduce costs and develop practical solutions to the unique challenges associated with Mars colonization. Mars colonization represents humanity's ambition, intelligence and ingenuity to establish a self-sustaining civilization in space. Although its cost could be astronomical, it is an investment that could yield significant scientific, technological, political and moral benefits. Pursuing a mission to Mars requires a bold act of faith, but it could also lead to one of the greatest achievements in human history.

EVALUATION OF SOURCES OF FINANCING

The cost of a manned mission to Mars is estimated to be between $100 billion and $500 billion, a staggering amount for any government or private entity to fund solely from internal sources. Although NASA has been exploring manned missions to Mars since the 1960s, the agency has fallen short due to both insufficient funding and lack of political support. In recent years, companies such as SpaceX and Mars One have entered the scene, offering a new dynamic for funding sources. Mars One, founded by Bas Lansdorp in 2011, announced plans to establish a permanent human settlement on Mars with a primary source of funding through a television program documenting crew selection and training. While the idea of publicizing the mission may seem controversial, Mars One has gained a significant following, demonstrating that there is a market for space and the public's desire to see this type of programming. With no guarantee that Mars One will succeed, critics worry that such an endeavor could be a "colossal waste of money for a vanity project" (Amos, 2018). On the other hand, SpaceX, founded by Elon Musk in 2002, has made significant progress toward colonizing Mars and plans to fund its mission with commercial space initiatives, such as satellite launches and commercial space travel plans. According to Musk, his goal is to make the cost of a trip to Mars roughly equivalent to the average price of a house in the United States, about $200,000, within the next few decades (Strauss, 2019). While undoubtedly still an expensive undertaking for the average citizen, it makes a manned mission to Mars more plausible and realistic. Critics of SpaceX argue that Musk's goals are overly ambitious and unrealistic, given the high failure rate of rockets in spaceflight. Ethical questions are raised about funding a Mars mission when the funds could be used to solve problems such as world hunger and poverty. Funding sources for Mars colonization must prioritize the needs of humanity while ensuring adequate support to make the dream of colonizing another planet a reality.

COSTS OF COLONIZING MARS vs. OTHER POSSIBLE SOLUTIONS

While many scientists and entrepreneurs believe that colonizing Mars is humanity's best chance of survival, others argue that there are less costly and more practical solutions that should be pursued first. One such solution is to focus on reducing greenhouse gas emissions on Earth, which could slow the pace of climate change and mitigate its worst effects. By some estimates, this could be achieved for a fraction of the cost of an initial Mars colonization mission, which is expected to cost billions of dollars. Investing in renewable energy sources could help reduce our dependence on fossil fuels and minimize the impact of resource depletion. Another possible solution is to focus on space technologies that could help us better understand and manage our own planet. For example, satellites and other remote sensing tools could be used to monitor global environmental changes and track the movement of critical resources such as water and crops. While these strategies may not be as exciting or glamorous as a mission to Mars, they could provide significant benefits for a fraction of the cost, making them a more practical and effective way to address the pressing challenges we face as a species. The decision to pursue Mars colonization will require careful consideration of both its potential benefits and its financial costs, along with a clear understanding of the other solutions available to us. The prospect of colonizing Mars has captured the imagination of scientists, policy makers and the general public. With overpopulation, natural resource depletion and global warming as existential threats to humanity, many see the red planet as a possible solution to ensure our survival. The idea of establishing a human settlement on Mars is not new. As early as the 1950s, scientists began to explore the possibility of sending humans to the planet. It has only been in recent years that a serious push to make this vision a reality has gained momentum. Technological advances, particularly in the field of space exploration, have made it possible to send unmanned missions to Mars, collect valuable data and knowledge, and pave the way for human exploration. Growing awareness of the fragility of our planet and the urgent need to preserve its resources has diverted attention to Mars as a possible alternative for sustaining human life. Proponents of Mars colonization argue that the planet offers many advantages over Earth. For starters, it has abundant water reserves, which can be extracted from its polar ice caps. The Martian atmosphere, composed mainly of carbon dioxide, offers the possibility of terraforming, a process of transforming an environment to make it more habitable for humans. This could involve, for example, pumping greenhouse gases into the atmosphere to raise its temperature and create a more Earth-like climate. Another advantage of colonizing Mars is its potential for scientific discovery and innovation. As humans explore the planet and conduct research, they will be able to learn more about its geological, chemical, and biological processes, as well as learn more about the history and formation of our solar system. This knowledge could have far-reaching implications for our understanding of the universe and the search for extraterrestrial life. The exploration and colonization of Mars could spur advances in technology and engineering that could have applications far beyond space exploration. For example, the development of a sustainable, self-sufficient habitat on Mars would require the development of new systems and technologies for food production, waste management, and energy generation, all of which could have applications on Earth. The challenges and risks associated with Mars colonization should not be underestimated. The journey to Mars alone poses significant logistical and engineering challenges, from developing spacecraft that can safely transport humans across the immense distance to designing habitats that can withstand the harsh Martian environment. Once on Mars, the challenges multiply. The Martian atmosphere is thin and lacks protection from cosmic rays, solar radiation and extreme temperature fluctuations. This poses a serious threat to the health of future colonists and may require the development of new technologies and medical treatments. The Martian environment is alien to human life, and colonists will have to adapt to living in a confined, isolated and potentially dangerous environment. There are also ethical considerations to be addressed. The question of who has the right to colonize another planet and what obligations it entails is complex and controversial. Some argue that Mars should be a common heritage of mankind, to be shared and managed collectively for the benefit of all. Others argue that Mars should be colonized by private companies or wealthy individuals, who will have greater incentives and resources to undertake this costly and risky venture. The potential for political, economic and social conflict between different groups of Mars colonizers is a real but as yet unknown possibility. Despite these challenges, the idea of colonizing Mars remains enticing, and advances in space exploration and technology make it increasingly feasible. It is important to recognize that Mars colonization is not a miracle solution to our planet's problems. It must be seen as one piece of a larger puzzle, along with efforts to address the root causes of overpopulation, natural resource depletion and global warming. It is important to approach Mars colonization with a responsible stewardship mindset, recognizing the ethical obligations that come with venturing into unknown territory. Only in this way can we ensure that Mars colonization does not repeat the mistakes of past colonialism and exploitation. In short, the challenge of colonizing Mars is not just a matter of technology or engineering, but of imagination, ethics and vision.

XI. RISKS AND THREATS

Despite the potential benefits associated with Mars colonization, it is crucial to examine the potential risks and threats associated with this endeavor. First, space exploration itself can be dangerous, and past missions provide clear examples of the risks involved, such as failed landings, communications failures, and fatal accidents. It is therefore essential to devote considerable time and resources to ensuring the safety of both the journey and the establishment of a colony on Mars. Creating a self-sufficient community on a planet so far from Earth would involve overcoming numerous technological and logistical challenges, such as developing sustainable energy sources and providing a reliable food supply. There is also the issue of psychological and mental risks, as prolonged periods of isolation can lead to depression, anxiety and other psychological disorders. The potential for social and political conflict cannot be ignored, as limited resources, such as food and water, can create competition and tensions between settlers. This could lead to a deterioration of the environment and threaten the success of the Mars colony.

Another potential risk involved in colonizing Mars is the danger of contaminating the planet with microorganisms from Earth. Although Mars is inhospitable to human life, it is possible that some microbial life exists beneath its surface or in its atmosphere. The introduction of terrestrial organisms, whether accidental or intentional, could pose a significant risk to any Martian life that may exist, and could result in irreversible contamination of the planet. Therefore, strict protocols must be established to ensure that any equipment, materials, or personnel sent to Mars are thoroughly sterilized, preventing the spread of terrestrial microbes on the dusty Martian surface. Any colonization effort must also be undertaken with a deep commitment to planetary protection, to ensure that human activity on Mars does not damage the planet's ecosystem, whether known or currently unknown.

In addition to the environmental and logistical risks, the establishment of a Mars colony would entail a high financial cost that would require significant investments in infrastructure and technology. The initial cost of Mars missions is already high: NASA's Mars InSight lander mission cost a total of $830 million. Establishing permanent habitation and maintaining a functional ecosystem on Mars would require far more resources, a figure that is currently beyond the reach of most governments and corporations. Adequate backing that could sustain the effort for decades or centuries would likely require an enormous amount of money and continued political support, and budget cuts that could come at any time could derail the endeavor. If a private entity funds this project, it raises the question of who governs the planet and raises profound ethical questions surrounding the privatization of Mars. Even if successfully established, a Mars colony would still be a precarious place with limited resources and the constant possibility of catastrophic failure. The harsh Martian environment presents numerous challenges, such as extreme temperatures, high radiation levels and dust storms that can last for months. Any technical failure could jeopardize the safety of the entire colony, and repair or replacement of equipment or supplies could be extremely difficult due to the distance and time delay involved in communicating with Earth. In the event of an emergency, there would be no possibility of evacuating colonists from the planet, further underscoring the need for effective interplanetary medical care and equipment development. The decade-long delay between the occurrence of a problem, the transmission of the situation to Earth, the possible planning and development of a response, and the dispatch of the tools, humans, robots, or materials needed to solve the problem could be catastrophic. Thus, even with continued support, developing and sustaining life on Mars will require a multigenerational commitment and the ability to withstand extreme challenges and unexpected emergencies. While Mars colonization may offer a promising solution to the global problems facing humanity, it is clear that the enterprise carries significant risks and threats that must be addressed. Safeguarding the safety and health of colony residents is critical, as is developing environmental and planetary protection protocols that ensure the sustainability of the Mars ecosystem. Funding, infrastructure, technological and medical advances, and political support will be essential to successfully establish and maintain a Mars colony. Humanity should carefully consider the potential costs before embarking on a journey to the red planet, taking into account the many risks involved, including possible irreparable damage to one of the few planets in our solar system that has not yet been explored by humans.

RISKS: PLANETARY PROTECTION, BIOLOGICAL CONTAMINATION AND SOCIAL UNREST

The prospect of Mars colonization brings with it a multitude of potential risks and threats that need to be addressed before moving forward. One of the main concerns is the impact it may have on planetary protection. The possibility of introducing new life forms or microorganisms to Mars could result in the contamination of the planet and the destruction of any possible indigenous life. This contamination could be accidental, as a result of human waste or microorganisms present on the surface from equipment used in the colonization process, or intentional, as part of scientific experiments or terraforming efforts. The International Committee Against Sample Return to Mars (ICAMSR) has expressed concern about the risk of returning samples from Mars to Earth and the possible contamination of the terrestrial biosphere. Biological contamination is another major risk associated with Mars colonization. The human body contains numerous microorganisms that are essential to sustain life on Earth, but these same microorganisms could pose a risk on Mars, where indigenous organisms are not adapted to them. The inability to contain all the human waste on the planet could lead to the spread of diseases and harmful bacteria that could have catastrophic repercussions on human health and the potential Martian ecosystem. To mitigate this threat, strict regulations and protocols are needed to prevent the spread of any potentially harmful microorganisms to the Martian surface. Aside from concerns about planetary protection and biological contamination, there is also the potential for social unrest to consider. The colonization process could give rise to a number of political problems, especially if large companies and nations compete for resources and territory on Mars. The possibility of exploitation of Martian resources could lead to significant socioeconomic and power struggles, with some groups being left behind or marginalized. This, in turn, could lead to conflict and unrest on both Earth and Mars. The psychological and social effects of being isolated from Earth and living on a planet with a different atmosphere and gravity could lead to mental health problems and strained social relationships. The prospect of colonizing Mars carries a multitude of potential risks and threats that must be addressed before moving forward. The impact it may have on planetary protection and biological contamination could lead to the destruction of any possible indigenous life and catastrophic consequences for human health and the potential Martian ecosystem. The potential for political and social unrest cannot be ignored, as the colonization process could exacerbate existing power struggles and lead to the marginalization of certain groups. To ensure that Mars colonization is carried out successfully and sustainably, strict rules, protocols and international cooperation must be employed, with a focus on safeguarding planetary protection, preventing biological contamination and mitigating social unrest.

HOW TO MITIGATE POTENTIAL RISKS AND THREATS

As promising as Mars colonization is, there are undoubtedly risks and threats that must be addressed for humanity to thrive on the Red Planet. One of the most pressing risks is the physiological and psychological risks that living in a low-gravity environment for prolonged periods of time can have on the human body. On Earth, gravity provides the necessary stimulation to our musculoskeletal, cardiovascular and vestibular systems, but these same systems can suffer significant atrophy when exposed to much lower levels of gravity. The isolation and confinement of living in a small, self-sufficient environment such as a Martian colony can lead to negative psychological effects such as depression, anxiety, and interpersonal conflict. To mitigate these risks, it will be necessary to develop new technologies and protocols that provide conditions as close as possible to Earth's gravity and environmental conditions. One solution could be to build larger habitable modules that allow for greater movement and physical activity and that can be pressurized to simulate Earth's atmospheric pressure. Frequent exercise and physical therapy will be essential to maintain bone and muscle density, and specialized "gravity bed" or "rotation bed" equipment may be needed to provide vestibular stimulation while allowing for restful sleep. There are also significant environmental risks that must be taken into account in any Mars colonization plan.

For one thing, Mars has limited resources, which means that any colony will have to be almost entirely self-sufficient to thrive.

With the potential for dust storms and adverse weather conditions, as well as the difficulty of extracting resources from the Martian soil, it will be essential to develop and implement sustainable practices, such as closed-loop systems that conserve and recycle water, air and waste. Energy production will also be a key factor, as Martian colonies will rely heavily on solar panels or nuclear reactors. The environmental impact of these energy sources and the safe disposal of nuclear waste will have to be carefully considered. The risk of contaminating the Martian environment with terrestrial microbes is a very real concern.

Strict protocols will need to be developed for the sterilization of equipment and other materials, as well as for the disposal of waste that may contain harmful bacteria or viruses. Beyond the physical and environmental risks, there are also important ethical considerations that must be addressed when it comes to the colonization of Mars. On the one hand, the colonial enterprise itself raises questions of ownership and property rights: Who owns the land on Mars and who has the right to exploit its resources? There is a risk that colonialism on Mars will mirror the exploitative practices of colonialism on Earth, with wealthy and powerful nations or corporations dominating and exploiting less developed colonists. To mitigate these risks, it will be important to establish clear rules and ethical guidelines that prioritize the welfare and autonomy of all Martian colonists, regardless of their social or economic status. International oversight and cooperation may be necessary in any Martian colonization effort.

There is a risk that colonization of Mars will exacerbate existing problems on Earth, rather than ameliorate them. Although sending overcrowded or resource-scarce Earth communities to Mars may bring short-term benefits, ultimately this solution does not

146

address the root causes of these problems. The resources and energy devoted to a Martian colony could be better spent on solving Earth's problems, for example, by investing in renewable energy technologies or addressing income and wealth inequality. To mitigate this risk, it will be important to ensure that Martian colonization is conducted in a way that enhances, rather than detracts from, ongoing efforts to address global challenges. This could involve redirecting resources and expertise from terrestrial industries to Martian industries, or developing technologies or practices on Mars that can be adapted and applied on Earth.

Overall, Mars colonization represents a major and complex undertaking, with a multitude of risks and threats that must be carefully considered and mitigated. With careful planning, collaboration and foresight, it is possible for humanity to establish a thriving and sustainable colony on the Red Planet. By addressing the physical, environmental, ethical and social risks associated with Martian colonization, we can ensure that this endeavor benefits our species and our planet as a whole, rather than harming them.

POTENTIAL RISKS AND THREATS vs. OTHER POSSIBLE SOLUTIONS

While colonization of Mars may present a promising solution to address the problem of overpopulation and natural resource depletion, it is not the only potential solution. Other solutions, such as technological advances in renewable energy, reduced carbon footprint and improved sustainable agricultural practices, among others, may prove to be less risky and more effective in the long term. The potential risks and threats associated with Mars colonization are numerous and may outweigh the benefits. The spread of communicable diseases, radiation exposure, and psychological problems are just a few of these potential risks.

First, the spread of contagious diseases is a major concern when it comes to Mars colonization. Because the Martian environment is so different from Earth's, the human immune system may not be able to handle the unknown viruses and bacteria that exist on Mars. If a potentially contagious disease were to reach Mars from Earth, it could spread rapidly throughout the colony and become uncontrollable. This could lead to deaths among the colonists and put an end to the colonization project. Second, radiation exposure is another potential risk associated with Mars colonization. The lack of a significant magnetic field on Mars exposes the planet's surface to solar and cosmic radiation, which can be harmful to humans and other living organisms. This radiation can cause genetic mutations, cancer and other health problems, making long-term exposure a serious concern. Since Mars has almost no atmosphere, any colonists on the Martian surface would be exposed to radiation doses similar to those received by astronauts while traveling through space.

Third, the effects that Mars colonization could have on mental health are also of concern. The isolation and confinement of living on Mars could have a considerable psychological impact on the colonists, causing high levels of stress and depression. The long mission duration and limited human interaction could lead to boredom, loneliness and even anxiety. Even with the most qualified and well-trained astronaut team, this type of isolation could negatively affect the mental and emotional well-being of the colonists. In analyzing these potential risks and threats associated with Mars colonization, it cannot be denied that this solution is extremely risky. This is especially evident when compared to other possible solutions to the problems of overpopulation and natural resource depletion. For example, a shift to renewable resources, such as solar, wind and hydroelectric power, could revolutionize our way of life. By reducing our dependence on fossil fuels, we could reduce our carbon footprint, reduce pollution and possibly even slow the effects of global warming. Sustainable agricultural practices that promote the use of organic fertilizers, terrace farming and crop rotation could reduce the need for pesticides, preserve soil quality and promote water conservation. Improved genetic engineering and the use of GMOs in agriculture could increase the world's food supply without requiring more land, water and resources. The possibility of creating drought-, flood- and pest-resistant crops could help protect food production and reduce the damage caused by natural disasters. Although colonization of Mars presents many risks and threats, it is undeniable that its potential benefits are great. Martian resources, such as minerals, water, and the possibility of harnessing geothermal energy, could contribute to the continued growth of human civilization. It is up to us to weigh the potential benefits against the risks and decide whether or not to continue colonization. It is also important to consider alternative, less risky solutions to the problems of overpopulation and natural resource depletion. By focusing on these solutions, we can mitigate the impact of human activities on Earth and take a step toward a more sustainable future. The idea of colonizing Mars has been gaining momentum in recent years as Earth's resources continue to deplete and overpopulation becomes an increasingly urgent problem. It is no secret that the Earth's resources are finite and, as the population grows, the demand for those resources will only increase. The depletion of natural resources, such as oil and water, has already led to conflicts and political tensions between nations. Global warming is causing irreversible changes in our planet's climate and poses a danger to all life on Earth. Driven by these environmental concerns, space agencies and private companies have set their sights on Mars as the next frontier. Mars, with its similarities to Earth, is seen as a potential site for human colonization, and the possibility of establishing self-sufficient colonies on the planet is seen as a potential solution to Earth's continuing and worsening ecological problems.

The concept of colonizing Mars is not new; the idea has been around for decades, but has always been seen as a distant and improbable possibility. Recent technological advances and the growing interest of governments and private companies have brought the possibility of colonizing Mars closer to reality. Exploration of Mars by rovers and other spacecraft has provided vital data and information about the planet, and it is now clear that there are many similarities between Earth and Mars. These include day length, seasonal patterns, and the presence of frozen water on its surface. These similarities offer hope that Mars may harbor life, and some scientists believe that the planet could even be terraformed to create a habitable environment for humans. The possibility of colonizing Mars raises a number of ethical questions. Is it right to use resources and funds to colonize another planet when there are still many problems on Earth that need to be addressed? Could colonizing Mars lead to the same environmental problems faced on Earth? There is also the question of whether it is ethical to send humans on a one-way trip to Mars, where they would face numerous dangers and uncertainties. Despite these concerns, the colonization of Mars is seen as a necessary step for the survival of mankind, and has generated a great deal of public enthusiasm and interest. One of the main goals of Mars colonization is to establish self-sufficient colonies that can survive without resources from Earth. This would require the development of advanced technology that can extract resources from Mars and convert them into usable forms. For example, companies such as SpaceX are developing reusable rockets that can carry people and equipment to Mars, and NASA is exploring ways to use the planet's resources to create fuel and oxygen. There is also the possibility of using 3D printing technology to create structures on the planet using local materials. These innovations are not only necessary for Mars colonization, but could also have applications on Earth, allowing us to live more sustainably and efficiently. Another advantage of colonizing Mars is that it could provide valuable scientific knowledge and insights into the origins and evolution of the solar system. Mars is the most Earth-like planet in our solar system, and studying it could provide clues about how planets form, the evolution of our planet, and even the possibility of life on other planets. Mars has already provided a wealth of information through exploration by robots and probes, and human exploration and colonization could provide even more detailed and nuanced data. Despite the challenges and ethical concerns, colonization of Mars is seen as a necessary step for human survival. With Earth's ecosystems and resources under pressure from overpopulation and global warming, the development of self-sufficient colonies on Mars could alleviate some of the pressure on our planet. Mars colonization would not be a definitive solution, but it could help us learn to live more sustainably and efficiently, both on and off Earth. Mars colonization could inspire future generations to engage in science and technology and provide valuable knowledge and information about our solar system and the universe as a whole. Ultimately, Mars colonization represents a bold and ambitious step for humanity, and a possible solution to our current ecological and environmental challenges.

XII. THE ROLE OF INTERNATIONAL COOPERATION

Successful colonization of Mars will require unprecedented international cooperation. No country can do it alone, and no nation should have exclusive control of the Red Planet and its resources. A collaborative approach would ensure that the best technology, experience and expertise are shared and utilized, leading to a more efficient and cost-effective mission. Collaboration among different nations and institutions would create a sense of shared ownership and responsibility for mission success, reducing the risk of conflict and competition. International cooperation in the colonization of Mars would also help to solve some of the ethical, legal and social problems that will arise. For example, who should be allowed to go to Mars and who should have access to its resources? How to ensure respect for the rights of the planet's indigenous life forms? What consequences will the terraforming of Mars have for its ecology and climate? These questions cannot be answered by a single nation or institution. Rather, they require a global conversation and collective action. International cooperation in Mars colonization would pave the way for more ambitious space missions in the future. Lessons learned from collaboration in Mars colonization could be applied to other space exploration ventures, leading to scientific and technological breakthroughs and new economic opportunities. A broader vision of space exploration, conceiving it as a collaborative effort, would also foster a sense of shared destiny for humanity. International cooperation in Mars colonization will not be easy to achieve. National interests, geopolitical rivalries, and different cultural and political systems could pose significant challenges. The history of international collaboration in space exploration has been uneven. While the International Space Station is a shining example of successful international cooperation, other initiatives, such as the Lunar Treaty, have not achieved universal support. To overcome these challenges and achieve successful international cooperation in Mars colonization, several steps are necessary. First, there must be strong political will on the part of all potential partners. Leaders must be willing to put aside their national interests and collaborate for the greater good. Second, a common goal and shared understanding of the benefits and challenges of Mars colonization must be established. It is important that all partners have the same vision and goals for the mission. Third, a legal framework must be developed that addresses the ethical, legal, and social issues of Mars colonization. A treaty that defines the rights and responsibilities of each partner and ensures the protection of the planet's ecology and indigenous life forms is essential. International cooperation in Mars colonization would require a significant investment of resources and funding. A diversified funding model that includes public and private investments, as well as international grants and donations, could help ensure the financial sustainability of the mission. Effective communication and collaboration among all partners would be key to the success of international cooperation in Mars colonization. Coordination mechanisms, such as regular meetings and joint decision-making bodies, could help facilitate communication and build trust among partners.

Information sharing and knowledge transfer mechanisms, such as joint research programs and training exchanges, could help build capacity and foster innovation. The role of international cooperation in Mars colonization is critical. Collaboration among different nations and institutions would provide the resources, knowledge, and expertise necessary for mission success.

International cooperation in Mars colonization would help address some of the ethical, legal, and social issues that will arise and pave the way for more ambitious space missions in the future. Achieving international cooperation in Mars colonization will require strong political will, a common goal, a legal framework, significant investment of resources, and effective communication and collaboration among all partners.

IMPORTANCE OF INTERNATIONAL COOPERATION

The future of humanity depends on our ability to explore new frontiers, and Mars represents a new opportunity to continue our quest for progress. In the face of overpopulation, resource depletion and global warming, the colonization of Mars appears to be our only hope for ensuring our survival. The success of this monumental goal depends on the cooperation of countries around the world. International cooperation is crucial to the success of Mars colonization efforts, as it requires an enormous amount of resources, technology and expertise that no single nation can provide alone. First, the cost of colonizing Mars is exorbitant, and a single nation cannot shoulder the burden alone. A key aspect of successful colonization is to ensure a continuous supply of resources to the colonists, who will be isolated from Earth for months or even years. This requires the development of complex transportation, communication and resource extraction infrastructures. Financial investment on such a large scale can only be achieved by pooling resources from several countries. Financial investment can be obtained from countries with a strong interest in space exploration, such as the United States, Russia, China and Japan. International cooperation allows nations to leverage their technological and financial resources to support each other in achieving a common goal.

Second, the technical expertise required for Mars colonization is beyond the capabilities of any single nation. Successful colonization will require a myriad of expertise, from rocket propulsion to deep space navigation and life support systems. By pooling the intellectual and technical resources of different countries, we can share our knowledge and learn from each other's experience. The end result will be faster progress and a greater likelihood of success. International collaboration will help ensure that the best minds from around the world are working toward a common goal with the sole objective of success. Third, the logistics of an undertaking as colossal as Mars colonization are beyond what a single country can provide. A multinational team will be needed to design, build and operate a Mars colony from scratch. Cooperation and collaboration between countries with different cultures and traditions will result in an ideal environment for the exchange of ideas and innovation, leading to new solutions to obstacles that might at first have seemed insurmountable.

International cooperation will foster more meaningful opportunities for the development of sustainable technologies that can be used to explore and colonize Mars. We know that any mission to another planet will generate waste, and we must instantly find ways to manage the accumulation effectively. The delicate ecosystem of Mars will not be able to support waste generation in the same way as Earth. With international cooperation, we can explore sustainable technologies that will allow us to live on other planets with minimal environmental impact. International cooperation in Mars colonization will strengthen global unity and collaboration, increasing the mutual interdependence of countries and fostering closer cooperation in other areas of international importance. The need for global collaboration in Mars colonization will attest to the level of importance that humanity and nations attach to the future of mankind and technological advancement. The Mars colonization revolution will be a capital success that no single nation will be able to manage alone, which will push the boundaries of what nations can achieve through cooperation. In short, the success of Mars colonization will depend on the availability and willingness to share sources, technology and knowledge among nations. No country could unilaterally create a self-sufficient colony on Mars. Through collaboration, the potential grows exponentially when a multinational team of astronauts, engineers, and other technical personnel can leverage the shared resources of member nations. In this sense, the colonization of Mars will be a shared achievement and a testament that cooperation among nations can accomplish extraordinary things. The world could jointly benefit from the knowledge gained and progress achieved through this bold and ambitious vision. Therefore, the importance of international cooperation in Mars exploration cannot be overstated.

CONFLICTS BETWEEN COUNTRIES IN EFFORTS TO COLONIZE MARS

As the race to colonize Mars accelerates, it is not hard to imagine the potential conflicts that could arise between countries. The allocation of resources, such as water and minerals, will undoubtedly be a point of contention. Historically, conflicts over resources have played an important role in human conflicts, and there is no reason to believe that the same will not occur on Mars. It is also possible that countries will vie for control of particular landing sites or areas of the planet where they believe they can more easily establish colonies. As with any new frontier, the gold rush is likely to drive countries to compete fiercely for territory and resources. The question is whether this competition will lead to violence or whether countries will find ways to work together for mutual benefit. There are a few reasons to be optimistic about the potential for cooperation. First, the cost of going to Mars is so high that it is unlikely that any one country could afford it on its own. By pooling resources and expertise, countries have a better chance of successfully launching a mission. Any successful mission to Mars will require a long-term commitment, which means that countries will have to work together to create sustainable colonies. There is no way for one country to establish a self-sustaining colony on Mars without the cooperation of others. There is a legal framework that regulates human activity in outer space, including Mars. The 1967 Outer Space Treaty, ratified by more than 100 countries, prohibits the militarization of space and explicitly states that no country can claim sovereignty over any part of space, including celestial bodies such as Mars. There is also reason to be concerned about potential conflicts. The space race between the United States and the USSR during the Cold War was driven in part by national pride and fear of being left behind technologically. It is possible that something similar could happen with the colonization of Mars, with countries competing for the prestige of being the first to establish a permanent colony on the planet. This competition could lead countries to cut corners on safety and environmental regulations, which could have catastrophic consequences. The legal framework for regulating human activity on Mars is largely untested. Although the Outer Space Treaty provides a general framework for regulating activity in space, it does not cover all contingencies that may arise during a Mars colonization effort. As a result, there may be disagreements between countries about what is and is not allowed under the treaty. Another area where conflicts could arise is terraforming. Terraforming is the process of modifying a planet's environment to make it more hospitable to human life. This would be a necessary step in creating a self-sustaining colony on Mars, as the planet's current environment is inhospitable to human life. Terraforming a planet can cause significant environmental damage and take centuries to complete. Different countries may have different approaches to terraforming, and some may be more willing to take risks than others. This could lead to disagreements and conflicts over how quickly to terraform the planet and how much environmental damage is acceptable. There is the question of intellectual property. Technologies needed for Mars colonization, such as new rocket engines and life support systems, are likely to be developed by private companies. These companies will try to profit from their inventions, which could lead to disputes over intellectual property rights. Countries may try to protect their own companies by implementing protectionist policies that limit access to foreign technologies. This could slow progress in Mars colonization and lead to tensions between countries. In summary, while there are reasons to be optimistic about the potential for cooperation in Mars colonization efforts, there are also areas where conflicts could arise. Resource allocation, the question of territory, concerns over safety and environmental damage, disagreements over legal frameworks, and disputes over intellectual property could cause conflicts between countries. It is important that countries work together to establish clear rules and guidelines for Mars colonization, and approach the enterprise in a spirit of cooperation rather than competition. The success of Mars colonization will depend on the ability of countries to work together toward a common goal.

INTERNATIONAL COOPERATION vs. OTHER POSSIBLE SOLUTIONS

As humanity looks toward the colonization of Mars as a possible solution to our current problems, the question of international cooperation arises. Is it effective in this endeavor and, more importantly, can it lead to the successful colonization of Mars? To answer these questions, it is necessary to compare international cooperation with other possible solutions. The first alternative solution worth examining is that of private companies. In recent years, private companies such as SpaceX and Blue Origin have emerged as major players in the space exploration industry. Their goal is also to colonize Mars, but their motivations and resources differ from those of governments. Private corporations are known for their efficiency and innovation, but in the case of a Mars colony, their drive for profit may hinder their efforts to work toward a common goal. On the other hand, governments have the power and resources to mobilize a large workforce and fund the necessary research and development. Differing national interests and bureaucratic hurdles can slow progress. As seen in the International Space Station project, which was a successful international collaboration, government cooperation can lead to progress. Even in this case, the project was largely U.S.-driven, which may limit the potential for a truly multilateral space enterprise. Another possible solution is a purely national perspective, with a single country embarking alone on Mars colonization. While this could eliminate the challenges posed by differing national interests, such a project would be incredibly resource

167

intensive and may not be feasible for a single country. In comparison, international cooperation would combine the resources and expertise of multiple countries, increasing the chances of success. The comparison shows that, although each solution has its merits, international cooperation appears to be the most promising approach in terms of maximizing resources and meeting the technological challenges of Mars colonization.

To maximize resources and meet the technological challenges of Mars colonization, further examination may be needed to determine the optimal approach to international collaboration in Mars colonization. The idea of colonizing Mars as a means of protecting humanity's future is not new. Since the advent of space exploration, scientists and science fiction writers alike have explored the potential of Mars as a second home for humanity. In recent years, this idea has gained more traction as the challenges facing our planet have become increasingly pressing. Overpopulation, depletion of natural resources and global warming have led humanity to desperately seek solutions to ensure its survival. For many, hope for the future lies in the colonization of Mars. The idea is not without its critics, who point to the immense technological, financial and political challenges that such an undertaking would entail. Proponents of Martian colonization argue that the benefits in terms of scientific advances, environmental sustainability and human survival justify the risks.

The question remains whether we should place all our hopes on Mars and neglect the pressing environmental problems on Earth itself, or whether we should pursue both options simultaneously. The decision whether or not to colonize Mars is complex and requires careful consideration and weighing of multiple factors.

XIII. CONCLUSIONS

Humanity has reached a critical point in its existence where resources and space are becoming increasingly scarce. Overpopulation, depletion of natural resources and global warming have contributed to these massive problems, and there is a real danger that, without swift action, the consequences could be dire. There remains the hope of the possible colonization of Mars, which offers a solution to these problems by providing a new frontier for human growth and expansion. Although still in its infancy, the technology continues to advance and develop, and experts predict that the day when humans can live and thrive on Mars may not be far off. It is essential that rapid progress is made and that all responsible parties work together to ensure that the necessary infrastructure is in place for such a feat. This involves cooperation between governments, companies and academic institutions, all working to provide the funding, research and technology needed to make Mars colonization a reality.

It is up to humanity itself to forge its future and its destiny, and if we are able to take the necessary steps to successfully colonize Mars, we will be able to ensure our existence and guarantee our survival for generations to come.

ENSURING HUMAN SURVIVAL: THE NEED FOR SOLUTIONS

As the world's population continues to grow at an unprecedented rate and humanity faces the twin challenges of environmental degradation and natural resource depletion, the need for urgent solutions to ensure our survival has never been more pressing. One key strategy advocated by many scientists and policy makers is the colonization of Mars. While the prospect of sending humans to colonize another planet may sound like science fiction, recent advances in technology and space exploration have made this a realistic possibility. Even if we succeed in establishing a colony on Mars, this alone will not be enough to ensure our long-term survival. We must also take decisive action on root causes of these pressing global challenges. This includes reducing our dependence on fossil fuels and transitioning to more sustainable forms of energy, investing in renewable resources and technologies, and implementing policies that support a more equitable and sustainable global economy. At the same time, we must also work to mitigate the effects of climate change, protect biodiversity and ecosystem health, and support the development of sustainable food systems that prioritize biodiversity, cultural traditions and the well-being of both humans and nonhuman animals. The future of humanity depends on our ability to come together to address these threats in a coordinated and effective manner. Only through collective action and a sustained commitment to sustainability can we ensure our survival for generations to come.

COLONIZING MARS TO ENSURE MANKIND'S SURVIVAL

One possible solution to ensure humanity's survival in the midst of global crises such as overpopulation and depletion of natural resources is the colonization of Mars. Mars offers a unique destination for space exploration, as it is the only planet in our solar system that closely resembles Earth in terms of potential habitability. Its proximity to Earth makes it a feasible location for space explorers, as a round trip to Mars would take approximately two to three years. Mars has valuable resources such as water and minerals that can be found on its surface, which would help maintain a self-sufficient colony. Establishing a human settlement on Mars could help alleviate Earth's population crisis by giving humans the opportunity to move to a new planet and start over. The colonization of Mars could lead to scientific breakthroughs and new technologies that would benefit all of humanity. Considering the potential benefits of colonizing Mars, it is important to critically evaluate the potential risks associated with such an undertaking, as well as the ethical implications of colonizing another planet. If done responsibly and thoughtfully, Mars colonization may be a viable solution to ensure the survival of humanity. The idea of colonizing Mars has been around for decades, but as the problems of overpopulation, natural resource depletion and global warming become more urgent, its importance is growing. Colonization is not the solution to the problems facing humanity, but rather a temporary fix to the apparent

173

difficulty for human survival. The mission to Mars is an exciting undertaking that captures the public imagination and inspires people to dream of possibilities for life on another planet. It is wrong to assume that Mars colonization can solve Earth's problems and become a substitute for traditional life. Colonizing Mars is a complex and costly endeavor that requires global cooperation, meticulous planning, and extensive engineering. Even if we succeed in getting humans to and from Mars safely, there are significant challenges to long-term sustainable habitability on Mars. These include obtaining sufficient energy for life, minimizing radiation impact, designing a water and food collection and distribution system, and creating a protected and viable habitat on an often unforgiving planet. Establishing a Martian colony will require not just a team of astronauts, but an entire community with diverse skills that can work together in synergy. This diversity of talent will be costly to create, and can only emerge from a very small subset of the population with the necessary knowledge and experience in highly skilled technological areas. Even if we succeed in establishing a colony on Mars, we must realize that the planet is not a substitute for Earth, and that it remains an inhospitable place to live. The climate of Mars is cold and dry, with little atmosphere and high levels of radiation. The resources available on Mars, such as water and oxygen, must be extracted from the planet itself or generated from limited atmospheric CO_2 reserves. The harsh conditions on Mars offer no possibility for cultivation, and the soil itself is contaminated with perchlorates, which are toxic to humans. Thus, the Martian colony must rely on Earth for almost everything it needs to survive, which in turn places an enormous economic burden. There is also the question of ethics and social responsibilities associated with Mars colonization. Who will go to Mars and who will stay on Earth? If we consider communities living on an unsustainable planet, shouldn't the possibility of life on Mars be shared with all humans equally? How do we prioritize funding for essential resources needed, such as healthcare, food, and housing for Earth's inhabitants, when we are spending an enormous amount of resources on exploring Mars? Is that expenditure justified in light of the many problems we face on Earth? It may be ethically uncomfortable to rationalize the prospect of colonizing Mars in this scenario. It is important to note that many of the problems facing humanity can be solved through efforts to create a sustainable way of life on Earth. To address overpopulation, we must continue our efforts to curb population growth, such as creating public health initiatives and expanding education for women. To minimize resource depletion, we must adopt a circular economy model that emphasizes the reuse and recycling of materials and energy. To mitigate the effects of global warming, we must focus on adopting renewable energy sources and reducing greenhouse gas emissions. These are all major challenges that require collective global collaboration, but they are achievable through effort and investment. While the idea of colonizing Mars is exciting, it is not a panacea for the many challenges facing humanity. Instead of looking to another planet and speculating on dreams of extraterrestrial habitation, we must face our problems and work to find sustainable solutions here on Earth. This will require effort, investment and cooperation from governments, businesses and individuals around the world. The colonization of Mars should not be a single goal that detracts from efforts to solve Earth's problems. On the contrary, we must take all the lessons learned from Mars exploration and invest them back into our planet. The only way to ensure our survival as a species is to accept the challenge of creating a sustainable future for ourselves on our planet.

XIV. IMPLICATIONS AND FUTURE OBJECTIVES

The prospect of colonizing Mars has profound implications for humanity. First, it represents an opportunity to mitigate some of the most pressing threats to our survival. By establishing a human presence on Mars, we can begin to alleviate the pressure that overpopulation, natural resource depletion, and global warming have imposed on our planet. Mars colonization can spur scientific discovery and innovation in ways we cannot yet foresee. For example, there is the potential to learn valuable lessons about how to create sustainable ecosystems that can support human life under adverse conditions. The development of new technologies and systems to sustain human life on the Red Planet may have broad applications in fields such as medicine, engineering, and ecology. The colonization of Mars also involves potential risks. Any undertaking of this scale requires significant resources and investment, and there are no guarantees of success. The harsh conditions on Mars can pose unforeseen challenges, and the long-term effects of living in a new and unfamiliar environment are difficult to predict. There is a risk that the resources and energy devoted to colonizing Mars will divert attention, resources and funding from more pressing global problems such as hunger, poverty and disease. Despite these risks, the prospect of colonizing Mars is an important and exciting one for humanity. It represents an opportunity to explore new horizons, push the boundaries of science and technology, and address some of the most pressing problems facing our planet. To realize the full potential of this enterprise, we must approach it with a clear assessment of its opportunities and risks, and with a commitment to use our scientific and technological capabilities for the betterment of humanity. We must also recognize that our success will depend on international cooperation, shared resources and a commitment to promote the common good.

Looking ahead, there are a number of questions we need to consider as we begin to explore the possibilities of colonizing Mars. For example, what are the specific scientific goals and objectives of a mission to Mars, and how can we design missions that balance scientific exploration with the practical challenges of sustaining human life on the Red Planet? How can we ensure that international collaboration and cooperation are at the heart of any mission to Mars and that the benefits of this enterprise are shared by all of humanity? What ethical and moral considerations should guide our approach to colonizing a new planet and how can we ensure that our efforts are based on a careful assessment of their potential impact on the environment, human health and safety, and broader social and economic implications? The colonization of Mars presents both great opportunities and significant risks to humanity. While the challenges of sustaining human life on the Red Planet are formidable, the potential benefits are numerous and far-reaching. With careful planning, international cooperation, and a commitment to the common good, we can realize the promise of this endeavor and push the boundaries of science and technology in exciting new directions. We must also recognize that the success of a mission to Mars depends on our willingness to approach it with humility, to learn from our past mistakes, and to work together toward a better future for all humanity. As we look to the stars and consider the possibilities of a new frontier, let us take on the challenges ahead with courage, optimism and a deep commitment to the values and principles that guide our efforts to build a better world.

IMPLICATIONS FOR THE GLOBAL COMMUNITY AND THE ECOSYSTEM

The colonization of Mars presents immense implications for the global community and ecosystem, as it represents a new step in human evolution. The colonization of Mars is not only a major milestone in scientific achievement, but also offers economic opportunities and optimism for the future. These activities could also have negative environmental repercussions on both Earth and Mars. The first potential implication of Mars colonization is the emergence of a possible shift in economic power, especially for the countries that succeed in colonizing the planet. If one or more countries become the first to initiate a sustainable colony, they will have an unprecedented economic advantage over all other countries and a significant redistribution of wealth would likely occur. Mars colonies could be self-sufficient and possess an abundance of valuable minerals, water and energy that could be commercialized. Successful colonization of Mars could also lead to revolutionary advances in agriculture, medicine, and industrial technologies, among many others, resulting in an increase in innovative companies and products. This new economic and social prosperity will undoubtedly give rise to new political realities; space colonization would redefine the challenges facing humanity. Countries would compete for control and access to the planet's resources, which could lead to interplanetary conflict. The colonization of Mars would have ecological and biological implications. Among them, the possible introduction of terrestrial microorganisms into the virgin Martian

181

environment. Prior to terrestrial explorations of Mars, the planet had no indigenous macro- or microflora or fauna. Exploration and colonization of Mars could introduce harmful pathogens to Mars, which would damage nanobiomes and disrupt natural processes that have been functioning for billions of years. As a consequence, Mars could be threatened by invasive species. Thus, the implications of Martian colonization on the planet's natural ecosystem are significant. However, the consequences of Mars colonization, whether positive or negative, remain largely speculative at this time; advanced measures and strict proce-dures need to be developed to reduce the various impacts of human activities on the planet. The colonization of Mars could also have positive repercussions on Earth in terms of environ-mental sustainability. A Mars colony could become a model for sustainable practices on Earth, as it would force humans to re-duce their environmental footprint. Sustainability is critical to maintaining the self-sufficiency of the Mars colony and making the most of limited resources. Planning actions on Mars would focus on building solutions that use fewer resources and produce less waste. The use of renewable energy sources on Mars would not only sustain human life on the planet, but also accelerate the global clean energy revolution on Earth. Colonization of Mars would provide humans with the opportunity to actively study the mechanics of sustainable settlements, which could be replicated on Earth. As a result, sustainable practices associated with Mars colonization could provide a pathway to address current envi-ronmental problems, slow the pace of environmental deteriora-tion, and promote long-term ecological conservation.

The colonization of Mars could have implications for the techno-logical advancement of humanity. Humans would have to create an entirely new environment on Mars through hydroponics and other life support systems, which would require a significant level of technological advancement and innovation. The development and application of advanced technology, improved farming systems, transportation and communication protocols, among other things, will be necessary for Mars colonies to remain self-sufficient. Technological advances developed for Mars colonization could have important benefits for technology as a whole. For example, the development of new technologies to better understand the surface of Mars could contribute to the development of better geological tools and the discovery of new resources on Earth. Research conducted on Mars could also lead to new advances in science and technology, including medical advances. The colonization of Mars, like any other major human endeavor, carries with it a number of ethical, social and environmental implications. The potential benefits of Mars colonization are considerable: scientific advances, economic prosperity, and environmental sustainability. Conversely, the potential dangers of Mars colonization include the introduction of destructive microorganisms from Earth into virgin Martian terrain, a possible threat to global security posed by the emergence of new space powers, and the inadvertent alteration of the Martian surface and geological processes. Although the possibility of colonizing Mars poses a unique challenge to mankind, it is of utmost importance that these implications be taken into account and the steps to be taken carefully planned before proceeding. In this way, humanity will be able to ensure that Mars colonization is carried out responsibly and with the interests of both Earth and Mars in mind.

FUTURE OBJECTIVES

The future of Mars colonization may be wide and varied, and offers great opportunities for the various actors interested in this topic. One possible approach to Mars colonization could be the development of self-sufficient colonies on the red planet, powered by solar energy and based on resources extracted from the Martian environment. This approach would be very ambitious and would require significant technological advances and investment in human resources to ensure that the colonies could be established and sustained over the long term. Another possible avenue for exploring Mars could involve sending manned missions to the planet, with the goal of conducting scientific research to help us better understand Mars and, potentially, to identify new avenues of exploration. These missions could span different scientific disciplines, such as geoscience, astrophysics and biology, and could also help pave the way for future colonization efforts by establishing a deeper understanding of the Martian environment. Alternatively, some researchers have proposed the establishment of a more commercial approach to Mars colonization, with the goal of creating new opportunities for exploration and development in space. This could involve the establishment of a manufacturing or research base on Mars, with the goal of providing a platform for private companies and other interested parties to explore the planet's potential for human habitation and resource extraction. This line of thinking is supported by the belief that Mars has significant potential as a source of valuable minerals and other resources, and that the development of a strong scientific and economic base on Mars could help drive innovation and stimulate economic growth in other areas. At the same time, there are also concerns about the potential impact of Mars colonization on the planet and its inhabitants, as well as on the broader social and environmental contexts in which these activities take place. For example, there are concerns about the potential impact of human activities on the Martian environment, including possible resource depletion or pollution, and the possible introduction of new species or microbes that could affect the Martian ecosystem. Also of concern are the ethical implications of Mars colonization, including issues of ownership, governance and sovereignty, as well as the potential impact of colonization on indigenous communities and other vulnerable populations. Overall, the possible future directions of Mars colonization are complex and multifaceted, requiring careful consideration of the various factors at play and the potential consequences of different approaches. Although the exploration of Mars and the possible establishment of colonies on the planet offer significant opportunities, it is essential that we approach this topic with caution and foresight, taking into account the many challenges and risks inherent in any such endeavor. The success of Mars colonization will depend on our ability to balance scientific advances, economic opportunity and ethical responsibility, and to adopt a sustainable and equitable approach to space exploration that benefits all of humanity.

FUTURE IMPLICATIONS vs. OTHER POSSIBLE SOLUTIONS

The idea of colonizing Mars has been seducing researchers and scientists for decades, but in recent years it has gained momentum and has become a more realistic solution to humanity's impending crisis. Other potential solutions have been proposed, each with its own implications and future directions. One of the most obvious solutions is to reduce human population growth, which would help alleviate pressure on our planet's resources. This can be achieved by various means, such as educating people, facilitating access to contraception and encouraging people to have fewer children. While this solution seems simple, its implementation is complex and requires a significant change in social norms. Another possible solution is to focus on renewable energy sources. Many countries have made progress in this field, and some are even aiming to become carbon neutral in the coming years. A shift to renewables such as solar, wind and hydro could significantly reduce our dependence on fossil fuels, which contribute to global warming and air pollution. Renewable energy sources are still in their infancy and require significant investment in research and infrastructure to become viable alternatives to fossil fuels. Similarly, the development of sustainable agricultural practices is another important solution to overcome the limitations of our planet. A shift toward regenerative agriculture can help restore damaged soil and reduce the harmful effects of conventional farming methods. Practices such as crop rotation, cover crops and intercropping can help improve soil

187

health and prevent soil erosion, which is essential to ensure sustainable food production. The idea of colonizing other planets such as Mars may be considered a long-term solution to the problems plaguing our planet. Since it involves building and maintaining human settlements on other planets, it requires technological advances that may not be feasible today. The prospect of becoming multi-planetary is exciting and would offer humanity a backup plan in case of catastrophe on Earth.

When comparing Mars colonization with other possible solutions, it is important to consider the implications and future direction of each option. Mars colonization is likely to have significant financial, technical, and ethical implications that must be taken into account. Although it offers a possible escape route for humanity, it is important not to view it as a quick or easy fix.

First, Mars colonization is a long-term solution that requires significant investments in time, money and resources. Our current technology is not advanced enough to sustain human life on Mars, and even with the most optimistic timelines, it will take decades to establish a self-sustaining human colony. The cost of such a mission is also expected to be enormous, with estimates ranging from tens of billions to trillions of dollars. This makes it a difficult problem to solve, given the current budgetary constraints of most countries. Second, the colonization of Mars has ethical implications. The idea of colonizing another planet raises questions about the ethics of space exploration and whether it is right to colonize other planets if we cannot take care of our own. There are also concerns about the safety of astronauts and the environmental impact of human settlements on Mars.

Therefore, it is important to consider the potential risks and benefits of colonizing Mars before committing to such a mission.

In comparison, other solutions such as reducing population growth, using renewable energy and developing sustainable agricultural practices have fewer ethical implications and can be implemented quickly. They also require significant changes in social norms, investments in research and infrastructure, and the political will to implement them. A comprehensive approach combining multiple solutions is therefore needed to address the complex challenges facing humanity. In terms of future directions, the colonization of Mars has the potential to transform our understanding of our universe and redefine what it means to be human. It offers humanity the opportunity to start anew and create new life on another planet. This future is not guaranteed and requires significant investments in research and technology to achieve it. On the other hand, solutions such as renewable energy and sustainable agriculture have a more immediate impact and can help ensure a more sustainable future for humanity on Earth. The decision to colonize Mars or to pursue other solutions depends on several factors, such as our priorities, resources and technological capabilities. It is important to remember that we have no choice but to act now. Humanity has reached a point where it must take bold and decisive action to preserve our planet's resources. Whether through the colonization of Mars or other solutions, the future of our planet depends on our ability to meet these challenges today. The idea of colonizing Mars has been circulating in the scientific community for decades, but in recent years it has gained considerable prominence due to growing concerns about overpopulation, depletion of natural resources and global warming. As the Earth's resources continue to dwindle and its climate becomes increasingly inhospitable, humanity is faced with the daunting task of finding alternative solutions to ensure its survival. The colonization of Mars has emerged as a possible solution to this problem. The Red Planet, which is the fourth planet from the Sun in our solar system, has long been considered the most promising candidate for human colonization due to its proximity to Earth, its relatively mild climate, and the presence of water and other resources that could be used to sustain life. The challenges posed by Mars colonization are considerable and will require significant investments in resources, technology and human capital. Some of the major challenges we will have to overcome are the high cost of sending humans and equipment to Mars, the harsh radiation environment, and the lack of a sustainable ecosystem on the planet. Despite these challenges, scientists and space agencies around the world are working tirelessly to develop the technologies, resources and strategies needed to successfully colonize Mars.

The colonization of Mars represents an exciting opportunity for humanity to explore new frontiers, create a new home for us beyond Earth and ensure our survival in the face of the pressing environmental challenges we face today.

XV. CALL FOR ACTION

With the challenges facing our planet in relation to overpopulation, depletion of natural resources and global warming, it has become clear that humanity must take critical steps to ensure its survival. The call to action that the need to colonize Mars implies is as urgent as it is necessary. It is an opportunity not only to ensure the continuity of the human species beyond our planet, but also to engage in groundbreaking scientific exploration and achieve unprecedented technological breakthroughs. As a global community, we must mobilize our collective intelligence, resources and technology toward this goal of establishing a permanent human presence on Mars. The journey to Mars is challenging, with significant technical, financial and logistical difficulties, but it is not an insurmountable task for mankind. The progress made so far in the field of space travel and exploration serves as proof of our ability to achieve seemingly impossible goals. This is a call to action for the world's governments, institutions, academics, businesses and individuals to unite and work towards the common goal of Mars colonization. We must strive to accelerate the current pace of innovation and development, creating new technologies to support the systems and infrastructure necessary for life on Mars. From life support systems to energy solutions to agriculture and manufacturing, we must develop self-sufficient systems to create a sustainable colony on Mars. We must develop self-sustaining systems that will enable the creation of a sustainable colony on Mars. Successful colonization of Mars will not only provide a solution to the problems facing our planet, but will also open up a vast world of possibilities for scientific exploration and discovery. The unique and unforgiving environment of Mars presents a perfect platform for diverse scientific investigations, which could lead to a much deeper understanding of our own planet and the universe beyond it. To achieve this, we must focus not only on the technical and scientific aspects of the mission, but also on the social and psychological challenges of establishing a new civilization.

We must plan for the social and psychological well-being of the colonists, taking into account the isolation and confinement they will face on Mars. Education and mental health support must be provided to ensure the mental and emotional stability of the first humans to live off Earth. Mars colonization should be approached as a global effort, emphasizing international cooperation, collaboration and diplomacy. Involving countries from all corners of the world in this mission can strengthen global ties, promote cultural diversity, and facilitate knowledge sharing among nations. Collaboration on the Mars mission would generate a level of international goodwill and cohesion that would support sustainable development here on Earth. Mars colonization should be seen not only as an opportunity to escape Earth's problems, but also as an enterprise that brings us closer together as a species. The call to action to establish a permanent human presence on Mars is a vital component of our efforts to ensure the survival of the human race. It is a journey that requires the investment of significant resources, the development of new systems, and unprecedented collaboration between individuals and nations. Successful colonization of Mars has the potential to usher in a new era of exploration and scientific discovery, but also to inspire and motivate the next generation of scientists, engineers and entrepreneurs. With a shared vision and a united global effort, we can transform this daunting but exciting mission into a reality, illuminating the potential to explore and innovate beyond our terrestrial boundaries.

CALL TO ACTION FOR INDIVIDUALS, PUBLIC AND PRIVATE ENTITIES TO SUPPORT MARS COLONIZATION EFFORTS

Overpopulation, depletion of natural resources and global warming are increasingly pressing problems, leaving humanity with no choice but to seek viable solutions to ensure its survival. Fortunately, the concept of Mars colonization holds great promise in this regard and is gaining traction among scientists, researchers and space enthusiasts. This colossal undertaking requires unprecedented support from individuals and public and private entities around the world. First, individual support for Mars colonization efforts is crucial in many ways. One of the most obvious ways to contribute is to advocate that NASA and other space agencies receive more government funding for their space exploration missions. This is a critical issue that is often overlooked and needs to be addressed. NASA's budget, for example, has been declining in recent years, from about 4.4% of the federal budget in 1966, during the Apollo program, to just 0.5% in 2020. This is a meager sum when the costs involved in colonizing Mars are taken into account. To put it in perspective, the Mars Rover mission alone cost a whopping $2.7 billion. Therefore, every voice counts when it comes to asking for more funding for space programs to help advance Mars colonization. Another way to contribute is to support nonprofit organizations such as the Mars Society, which advocates for Mars colonization. Its founder, Robert Zubrin, has stated that the Mars Society's main goal is to instill the vision of a human mission to Mars in the public mind. To this end, they promote Mars exploration with simulations, contests and educational initiatives. The greater the number of people who support these organizations, the greater the likelihood that they will be able to garner the resources and partnerships necessary to make Mars colonization a reality. Private companies such as SpaceX are taking the lead in Mars colonization efforts. They have made great strides in developing reusable rockets that have significantly reduced the cost of spaceflight. The company has also developed ambitious plans to send manned missions to Mars in the coming decades, and CEO Elon Musk estimates that they could send humans to Mars as early as 2026. Achieving such ambitious goals requires significant financial investment from the private sector. This is an area where governments, wealthy individuals and investors can play a crucial role in supporting private companies like SpaceX. Government incentives such as search and development grants, tax breaks, and public-private partnerships could spur greater private sector investment in Mars colonization efforts. If investors and wealthy individuals could form a coalition to bring more significant investment to private space ventures and offer their expertise in a variety of areas, such as engineering and materials science, they could further boost Mars colonization efforts. With sustained private sector investment, the cost of Mars colonies will decrease and the number of discoveries could increase, making the project a reality in a shorter time frame. Third, public support, particularly from world leaders and major organizations, is essential to create good governance for the Mars colonization project. This requires the development of international protocols around property rights, security, environmental concerns and cooperation. For this reason, the United Nations created the Committee on the Peaceful Uses of Outer Space (COPUOS) in 1959. COPUOS is responsible for developing international laws and guidelines for space exploration and has been instrumental in guiding the space activities of nations. This demonstrates that world leaders see the value of space exploration and are willing to work together to achieve its goals. The global community must continue to support and engage with COPUOS to provide unified strategic leadership and support the governance of the Mars colonization project. The colonization of Mars offers a potentially exciting and viable solution to the alarming problems facing humanity. But the dream of having humans on Mars within our lifetimes is not without significant challenges that require a concerted effort from all fronts. Governments, private companies and individuals have a unique role to play in advancing Mars colonization efforts. Individuals can engage in advocacy, both online and offline, supporting the work of non-profit organizations, while private investment can fund research and development and drive progress. Public support can provide a clear and enforceable policy framework to ensure project sustainability and success. The task ahead is enormous, but with the right support and resources invested in Mars colonization efforts, we can leverage the promise of space exploration to ensure the survival of our species.

PUBLIC AND PRIVATE INITIATIVES FOR MARS COLONIZATION

Public and private initiatives to colonize Mars have gained momentum in recent years. NASA, the European Space Agency (ESA), SpaceX and Blue Origin are among the leading contenders in the race to reach Mars. NASA's Mars 2020 mission, scheduled for launch in July 2020, aims to explore Mars' Jezero Crater to better understand its geology and search for signs of past microbial life. NASA has also launched the Artemis program, which aims to establish a sustainable human presence on the Moon, paving the way for eventual human missions to Mars. ESA has coordinated the successful ExoMars missions in 2016 and 2018, which aimed to search for bio-signs of life on Mars. SpaceX, founded by Elon Musk in 2002, has ambitious plans to build a self-sufficient city on Mars with up to one million residents. Musk aims to achieve this by establishing a self-sufficient settlement on Mars, which would produce all the resources necessary for survival, from food, water and oxygen, to the necessary building materials. Blue Origin, founded by Amazon CEO Jeff Bezos in 2000, has also shown interest in space tourism and plans to establish permanent human settlements on the Moon to pave the way for Mars colonization in the future. Private missions have also gained momentum in recent times. The Mars One mission, founded in 2012, aimed to establish a permanent human settlement on Mars by the mid-2020s. The mission attracted numerous applications from more than 200,000 people from around the world willing to undertake a one-way trip to Mars. Due to various operational and financial problems, the mission was abandoned in early 2019. Similarly, the Inspiration Mars Foundation, founded by billionaire entrepreneur Dennis Tito, intended to launch a flyby mission around Mars with a couple in the next decade. The mission was intended to take advantage of a rare window of opportunity in which Mars would be very close to Earth to minimize the duration of the trip. The foundation failed to attract the funding and support needed to execute the mission. The growing popularity of Mars colonization can be attributed to the planet's potential as a suitable location for human habitation. Despite its harsh environmental conditions, the planet has several advantages that make it an ideal destination for long-term human settlement. First, Mars has an Earth-like surface, with mountains, valleys, deserts, and polar ice caps, making it easy for humans to acclimate to their environment.

Second, Mars has a day-night cycle of approximately 24 hours and a seasonal cycle of about 687 Earth days, making it more feasible for humans to adapt to its circadian rhythms. Third, Mars has abundant frozen water resources in its polar caps and subsurface, which can be extracted and processed to provide a renewable source for drinking water, agriculture, and industrial applications. Fourth, Mars has a thin atmosphere, which poses minimal radiation risks compared to space, reducing health risks to astronauts. Mars also has a relatively large equatorial bulge, which offers the opportunity for rotating habitats to generate artificial gravity, essential for human health and well-being on long-duration missions. The search for solutions to ensure human survival on Earth has led to a growing interest in Mars colonization. Public and private initiatives towards Mars colonization have gained momentum in recent years, and organizations such as NASA, ESA, SpaceX and Blue Origin are among the leading contenders in the race to reach Mars. The potential for long-term human settlement on Mars is significant due to its unique advantages, such as an Earth-like surface, abundant frozen water resources, a thin atmosphere, and the opportunity to generate artificial gravity. The ability to establish a self-sufficient and self-sustaining settlement on Mars, producing all the necessary resources, holds great promise for the long-term sustainability of human civilization. Although several technical, operational and financial challenges remain to be overcome, Mars colonization offers a beacon of hope for securing a better future for humanity.

EMPHASIZING THE ADVANTAGES OF COLONIZING MARS TO SECURE HUMAN FUTURE

The colonization of Mars represents an opportunity for humanity's survival that cannot be ignored. By establishing a self-sustaining human settlement on Mars, humanity could diversify its risk factors and ensure that the fate of our species is not tied solely to planet Earth. In fact, some scientists believe that the long-term survival of our species can only be assured by colonizing other planets, such as Mars. The idea stems from the inherent risks to Earth from various natural calamities such as earthquakes, hurricanes and asteroid strikes that could render it uninhabitable. Overpopulation, natural resource depletion and climate change are becoming increasingly critical issues for our survival on Earth, and the colonization of Mars offers a possible solution to these problems. By establishing a Martian settlement, we could build an ecosystem designed to support human life and, at the same time, be able to provide resources that have been depleted or depleted on Earth. This could include resources such as water, minerals, and vital elements such as oxygen and nitrogen. The potential benefits of colonizing Mars for humanity's survival also include the ability to test our mettle as a species. By pushing ourselves to establish a settlement on a completely new planet, we would test our limits of technological innovation, which could open up opportunities for new developments in a variety of fields. This also represents an opportunity to come together as a global community and collaborate on a collective project that is larger than any one nation. Through the Mars colonization project, we could develop a shared vision for the future of humanity that is technologically advanced, sustainable and focused on preserving our species for centuries to come. The potential benefits of establishing a Martian settlement are immense, and we can no longer afford to ignore them. To do so would put us at risk of becoming extinct as a species due to factors beyond our control. Colonizing Mars provides us with the opportunity to forge a new path for humanity, one that embraces innovation, collaboration and sustainability, and one that ensures that our species will continue to thrive in the future.

BIBLIOGRAPHY

Jeffrey Eisenberg. 'Call to Action.' Secret Formulas to Improve Online Results, Bryan Eisenberg, HarperCollins Leadership, 10/29/2006

Raymond J. Halyard, 'The Quest for Water Planets.' Interstellar Space Colonization in the 21st Century, American Eagle Publications, 1/1/1996

Sven Erik Jørgensen. 'Global Ecology.' Academic Press, 4/16/2010

Peter David Blanck. 'The Americans with Disabilities Act and the Emerging Workforce.' Employment of People with Mental Retardation, AAMR, 1/1/199

Kendra R. Parker. 'The Bloomsbury Handbook to Octavia E. Butler.' Gregory J. Hampton, Bloomsbury Publishing, 2/20/2020

Maurice Blaise. 'White Mandingo Part II.' The Conclusion, iUniverse, 2/1/2001

John B. Charles. 'Human Health and Performance Risks of Space Exploration Missions.' Evidence Reviewed by the NASA Human Research Program, Jancy C. McPhee, National Aeronautics and Space Administration, Lyndon B. Johnson Space Center, 1/1/2009

Konrad Szocik. 'The Human Factor in a Mission to Mars.' An Interdisciplinary Approach, Springer, 4/9/2019

The Princeton Review. 'Princeton Review AP English Language & Composition Prep, 2023.' 5 Practice Tests + Complete Content Review + Strategies & Techniques, Random House Children's Books, 8/16/2022

Division on Engineering and Physical Sciences. 'Safe on Mars.' Precursor Measurements Necessary to Support Human Operations on the Martian Surface, National Research Council, National Academies Press, 6/29/2002

Adriana Cordali. 'Visual Rhetorics of Communist Romania.' Life Under the Totalitarian Gaze, Springer Nature, 1/10/2023

Howard Whitton. 'Managing Conflict of Interest in the Public Sector.' A Toolkit, Organisation for Economic Co-operation and Development, 1/1/2005

Andrea Sommariva. 'The Political Economy of the Space Age.' How Science and Technology Shape the Evolution of Human Society, Vernon Press, 2/28/2018

Robert Zubrin. 'The Case for Mars.' The Plan to Settle the Red Planet and Why We Must, Simon and Schuster, 2/2/2021

Antonio Viviani. 'Mars Exploration.' a Step Forward, Giuseppe Pezzella, BoD - Books on Demand, 9/9/2020

Tony Milligan. 'The Ethics of Space Exploration.' James S.J. Schwartz, Springer, 7/25/2016

Robert Zubrin. 'The Case For Mars.' Simon and Schuster, 12/11/2012

National Aeronautics and Space Administration. 'NASA's Journey to Mars: Pioneering Next Steps in Space Exploration.' Government Printing Office, 2/5/2016

Joseph Seckbach. 'Terraforming Mars.' Martin Beech, John Wiley & Sons, 11/18/2021

Viorel Badescu. 'Mars.' Prospective Energy and Material Resources, Springer Science & Business Media, 12/7/2009

William Sims Bainbridge. 'Goals in Space.' American Values and the Future of Technology, SUNY Press, 1/1/1991

Joseph N. Pelton. 'The New Gold Rush.' The Riches of Space Beckon!", Springer, 11/4/2016

Robert Zubrin. 'Case for Mars.' Simon and Schuster, 6/28/2011

John Noble Wilford. 'Mars Beckons.' The Mysteries, the Challenges, the Expectations of Our Next Great Adventure in Space, Knopf, 1/1/1990

Isecg. 'Benefits Stemming from Space Exploration.' DIANE Publishing Company, 10/24/2013

Division of Behavioral and Social Sciences and Education. 'Global Environmental Change.' Under- standing the Human Dimensions, National Research Council, National Academies Press, 2/1/1991

Chris Prophet. 'SpaceX from the Ground Up.' 4th Edition, Independently Published, 11/21/2017

Mecha Summarizer . 'Energy and Environmental Science: A Guide to Sustainable Solutions.' Individual Unfold , 4/14/2023

Intergovernmental Panel on Climate Change (IPCC). 'The Ocean and Cryosphere in a Changing Climate.' Special Report of the Intergovernmental Panel on Climate Change, Cambridge University Press, 5/19/2022

Mark Maslin. 'Global Warming: A Very Short Introduction.' Oxford University Press, UK, 11/25/2004

David E. Naugle. 'Energy Sprawl Solutions.' Balancing Global Development and Conservation, Joseph M. Kiesecker, Island Press, 6/15/2017

Sjak Smulders. 'Sustainable Resource Use and Economic Dynamics.' Lucas Bretschger, Springer Science & Business Media, 7/19/2007

André Marcel Diederen. 'Global Resource Depletion, Managed Austerity and the Elements of Hope.' Eburon Uitgeverij B.V., 1/1/2010

Micah Sanchez. 'Natural Resource Depletion.' The Rosen Publishing Group, Inc, 12/15/2017

Department of Economic and Social Affairs, World Social Report 2020. Inequality in a Rapidly Changing World, United Nations, 2/14/2020

Ann Foley Scheuring. 'Global Climate Change and California.' Potential Impacts and Responses, Joseph B. Knox, University of California Press, 1/1/1991

Rebecca Stefoff. 'Overpopulation.' Chelsea House Publishers, 1/1/1993

Foundation for Deep Ecology. 'Overdevelopment, Overpopulation, Overshoot.' Tom Butler, Foundation for Deep Ecology, 1/1/2015

Richard Wagner. 'The Case for Mars.' The Plan to Settle the Red Planet and why We Must, Robert Zubrin, Free Press, 1/1/1996

Leslie Sklair. 'The Anthropocene in Global Media.' Neutralizing the risk, Routledge, 11/22/2020